Printmaking in the Service of Botany

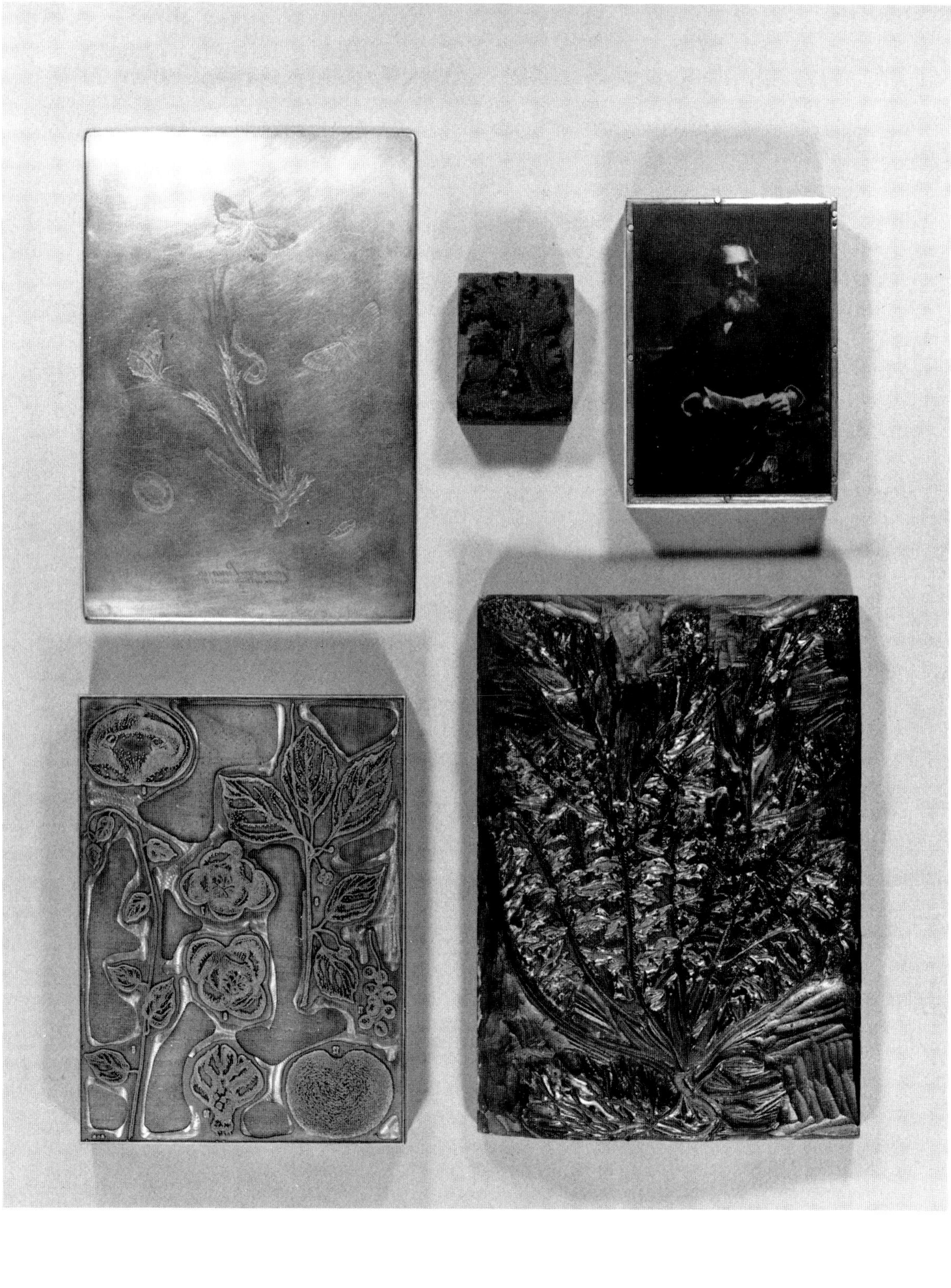

Printmaking In The Service Of Botany

21 April to 31 July 1986

Catalogue of an Exhibition

Gavin D. R. Bridson and Donald E. Wendel

with the assistance of James J. White

HUNT INSTITUTE FOR BOTANICAL DOCUMENTATION

CARNEGIE-MELLON UNIVERSITY

PITTSBURGH

1986

ISBN: 0-913196-49-5

Introduction

This is not an exhibition about the art of botanical drawing. It is, instead, a brief survey of the way that botanical drawing has got into print. The rôle of the artist in plant science is too well established to need further justification here. But the need for means of multiplying and disseminating accurate copies of his work has been almost as important to science as the need for the drawing in the first place. However simple or sophisticated his drawing was, it could only be of real service to science when botanists at large could study and use its information.

Before the advent of printing there had been a long history of illustrated botanical manuscripts, probably dating back to before Christ. The earliest that has survived is the stately *Codex Aniciae Julianae*, which dates from the early 6th century.* Unfortunately for science, very few people could make any use of these documents, let alone know of their existence. Of course, illustrated manuscripts could be duplicated, but it was an extremely slow and difficult task. Those who did make copies, and others who made copies of those copies, often debased the illustrations in the process, or even omitted them altogether.

The arrival of printing opened many new avenues to science, amongst them the possibility of multiplying images. Whereas a manuscript copy frequently differed from the original (though only the artist and anyone else who compared the two might know it), the printed picture ensured a fixity of the image, however good or bad it was. In an age when we hear public figures habitually complaining that they have been misquoted, we may appreciate the position of the manuscript artist when copyists distorted his visual statements. In the long term, printing was to provide extremely successful solutions to this problem. The path, though, was very uneven and many external factors affected its course, some beneficial, others retrogressive. As a result, botanical prints and illustrations range from the crude to the magnificent, from the patently experimental to the technologically sophisticated. In the long term botanical artists were very well served.

At its inception, however, this new technology, which was principally concerned with multiplying words, was slow to devise successful and economical methods for accurately reproducing graphic works. In the late 15th century a consummate artist, like Albrecht Dürer, might himself print images of outstanding beauty and complexity, but no ordinary printer could hope to employ craftsmen of that calibre to make book illustrations. Such workers scarcely existed, and the few who did were likely to be employed by wealthier patrons than book printers. In this respect the early botanical publications seem to have suffered rather badly. Despite the existence of contemporary botanical painting of commendable skill and accuracy, the earliest botanical book illustrations are very disappointing to our eyes and must surely have failed to please their contemporary readers also. Their typography, on the

*For a modern full-color facsimile of this manuscript see: *Dioscurides. Codex Vindobonensis Med. Gr. 1, der Österreichischen Nationalbibliothek*. Graz, Akademische Druck- und Verlagsanstalt, 1965-70, with "Kommentarband zu der Faksimileausgabe" by Hans Gerstinger.

other hand, is often praiseworthy, which leads one to conclude that early printers, by their nature, were far less interested in pictures than in words.

Over the last five centuries, botanical artists have employed a variety of media, such as pen and ink, pencil, charcoal, colored chalk, transparent or opaque watercolor, gouache and even oil paint, working on paper of various kinds, vellum, board or canvas. This range of materials has permitted a number of artistic styles, as well as a great latitude for personal expression within them. The many printing techniques devised over hundreds of years for reproducing this range of styles and mediums is too extensive to illustrate comprehensively here. The *Index Londinensis* lists some 487,000 illustrations of vascular plants published between the early 18th century and 1920. There is no available estimate of the grand total of botanical illustrations, which would include pre-18th-century and post-1920 depictions of vascular plants, as well as those of cryptogamic plants from all periods. This exhibition highlights 72 distinctive examples to outline the story of five centuries of printmaking in the service of botany.

The methods of printing pictures are usually divided into three classes: relief, intaglio and planographic. These terms relate to the printing surface rather than the printed image, which is always no more than ink on paper. A relief surface stands up a millimeter or so from the block or plate so that only the image touches the paper (Figs. A-D). For an intaglio surface the image is incised or etched into the face of a polished metal plate. When the plate is smeared with ink and superficially wiped clean, the incisions retain ink which, under great pressure, prints onto paper (Figs. E-G). A planographic surface is a flat, porous stone (or a grained metal plate) which carries a drawn or photographically created image that has an affinity with greasy ink. The rest of the surface, which holds water when wetted, rejects ink, so that only the image area prints (Fig. H).

A byway of printing, one almost entirely confined to botanical illustration, produced the so-called "nature prints." These processes utilized the plant itself to make the print or to create the form of the image on the printing surface, and they do not readily fit into the traditional classes of printmaking.

After four centuries of hand-produced printing surfaces, photography was discovered in the mid-19th century. In its infancy photography could produce only singly-developed images, but before long it was applied to graphic-arts printing. The combined agency of light and chemistry could replace handwork in the preparation of printing surfaces, and "photomechanical" techniques, like the older methods, were developed to create relief, intaglio and planographic printing surfaces.

The examples presented here are broadly arranged in these three classes, with some chronological and methodological subdivision. The arrangement is as follows:

RELIEF (EARLY)
Woodcut #1-5

INTAGLIO
Etching and line engraving #6-21
Tonal. Roulette and stipple engraving #22-25
Soft-ground etching #26
Aquatint and mezzotint engraving #27-30
Hand coloring #31-34
Intaglio color printing #35-39
Steel engraving #40-41

RELIEF (LATER)
Wood engraving #42-46
Relief color printing (chromoxylography) #47-48
Unconventional relief processes #49-50

PLANOGRAPHIC
Lithography #51-55
Planographic color printing (chromolithography) #56-58

NATURE PRINTING #59-61

PHOTOGRAPHIC
Cyanotype #62
Wood-engraved photogram #63
Photomechanical processes #64-68
Photomechanical color printing #69-72

The first items (#1-5) exemplify the earliest botanical illustrations. All are woodcuts, a medium which rose from crude beginnings in the later 15th century to a peak of excellence by the mid-16th, only to decline rapidly afterwards.

The later 16th century saw the earliest botanical use of etching, an intaglio process that was to remain fundamental to all linear printmaking methods for another four centuries (#6-11). Line-engraved botanical plates appeared soon after the first etchings and had an equally long history (#12-18). The two processes were often combined on the same plate. Although linear botanical prints were often hand-colored, some were specially designed to preclude the necessity for coloring (#19-21).

The 18th century introduced tonal methods of engraving such as chalk manner, stipple and mezzotint (#22-30). Of these, stipple engraving was by far the most successful for plant illustration.

The use of color in botanical printmaking is the next focus. By the end of the 18th century, hand coloring had grown to be an important adjunct to botanical engraving (#31-34). About that time, intaglio color printing was introduced with conspicuous success in botanical work (#35-40).

Relief processes were reintroduced for botanical illustration following the late-18th-century English revival in wood engraving. The next century saw great refinements in wood engraving, excellent relief color printing, and some experimental attempts to replace wood engraving (#42-50).

In 1811 the newly invented lithographic process was first used for botanical illustration, and it would serve the science for the rest of the century (#51-55). Its adaptation for color printing was particularly valuable (#56-58).

The byway of "nature printing," in which the plant was its own artist, is demonstrated (#59-61) before we turn to early photography (#62-63), in which, again, the plant created its own image. Plant photography from life was introduced in 1860, and we touch briefly on the early history of its reproduction by printing (#64-66). The general history of natural-history photography is a subject for a separate exhibition some day.

We close with a concise review of the rôle that photomechanical printing has played in reproducing botanical art—line drawing, tinted wash illustration, and various styles of watercolor painting, dating from the late 19th century to the present day (#67-72).

Each example is illustrated by photographs on three scales: a general view of the print, a section of the print seen in life-size, and an enlargement of the same detail. Our intent is to lead the eye toward a view of the *anatomy* of the print and to demonstrate features discussed in the text. The notes for each entry highlight the special aspects of the printing process that were significant for plant illustration. Taken as a whole, the series of notes is designed to have some historical continuity.

As previously noted, the classification of prints is usually based on the nature of the printing surface. A print is only ink and paper, perhaps with added watercolor. But the nature of the paper, the way the ink sits on its surface, and the method of application are features that contribute to the character of a print and are worthy of study. They are also of analytical value in determining the printmaking process.

The composition, malleability, surface texture and stability of the paper contribute much to the character and quality of a printed picture. All printing has been performed under pressure, and most paper, especially early handmade, is compressible. Printing has often relied on that quality, and printers usually have enhanced it by first dampening the sheet. Moist paper, when compressed under the immense force of a printing press, tends to remain compressed. The very term "impression" suggests this. A glance at any early book will show that each letter of the type has a sunken impression; one can also feel it. All relief printmaking processes have done the same, to some degree, though it may require magnification to see it.

Intaglio surfaces also compress the paper, but over the entire plate. The sunken "platemark" is a characteristic of the medium. The incisions that form the image obviously exert less pressure than the plate surface, so the paper is less compressed in those places. A "blind" (i.e. uninked) impression from an intaglio plate actually creates a minutely embossed and clearly discernible image (Fig. M).*

*For an example of a complete "blind" print see Plate 5 in Sir Frank Short's *Etchings and Engravings: What They Are, and Are Not...* London, Royal Society of Painter-Etchers and Engravers, 1912.

Planographic prints lack any of these surface sculptings. At most one can sometimes see a slight smoothness in the surface texture of older lithographs where the paper was forced against the stone by the intense pressure of printing.

Older printing ink was composed of a stiff "varnish" and a pigment. A relief surface, when printed on damp paper, forced some ink to ooze out at the edges of the lines or dots. This "squash" is often clearly visible under magnification. Intaglio ink had a specially viscous consistency, and dried on the paper with a faintly tangible thickness. Atop the embossed lines and dots of the image, it adds somewhat to their prominence. This slight elevation can be felt clearly with the fingertip on many intaglio prints.

The surface of most paper is not smooth when seen under high magnification. Very old handmade paper is especially rough and sometimes bears the texture of the felt interleavings between which it was pressed and dried during manufacture (Fig. I). Furthermore, old paper usually shows the wire pattern of the mould in which it was made. These parallel lines (called chain and wire lines) are what identify a "laid" paper when seen against the light. The pattern is due to differences in thickness of the paper and, under a raking light, shows in the surface texture of the "wire side" (Fig. J). From the late 18th century onwards, the wire on papermaking moulds was much reduced in gauge and woven into a fabric like mesh. "Wove" paper, produced in this kind of mould, shows no visible see-through pattern, though the texture of the wire side may reveal a pattern under a raking light (Fig. K). These features of surface texture—wire impressions, watermarks, felt marks and other irregularities—could adversely affect the quality of an impression and spoil delicate work. Printers were well aware of these hazards, and normally printed on the "top" rather than wire side of the paper. The introduction of smooth, even-textured machine-made papers in the 19th century eliminated some of the problems that beset earlier printers (Fig. L). It was also possible to add extra polish to the already smooth surfaces of these papers by a process known as "calendering." A greatly increased range of specialized papers became available, and the use of the correct kind became crucial to the success of sophisticated printing processes.

The photographic enlargements of the exhibits reveal some of the anatomical features that contribute to the character of the print: surface texture of paper, impression of relief line, elevation of intaglio line, ink squash, line profiles, shape of line ending, shapes and placement of dots, types of print grains, anatomy of coloring, and more. The accompanying notes will indicate what to look for in each magnified detail.

The Hunt Institute collections have provided most of the items in this exhibition, including some that are on indefinite loan from the Department of Botany, Smithsonian Institution. Additional material has been lent by Hans P. Kraus Jr., New York (#62), the Library of the Academy of Natural Sciences of Philadelphia (#64), and two anonymous lenders (#26, 63). Ancillary material has been lent by the Division of Graphic Arts, National Museum of American History, Smithsonian Institution (Prestele's lithographic stone, Smillie's aquatint-etched plate), Richard Brunkus, Albion, Michigan ("blind" printed etching), and Francis M. Hueber, Chevy Chase, Maryland (Curtis line- and stipple-engraved copper plate), to all of whom we are most grateful.

We thank James White, Jean Gunner and Bernadette Callery of the Hunt Institute for much help in assembling and preparing these materials for exhibition. For assistance in arranging loans for the exhibition we thank Janet Evans of the Library of the Academy of Natural Sciences of Philadelphia and Elizabeth Harris and Helena Wright of the Division of Graphic Arts, National Museum of American History, Smithsonian Institution. Frank Reynolds of the Hunt Institute made all the photographs after patient consideration of special problems and much experimentation with the enlarged details. Anita Karg, Charlotte Tancin and Michael Stieber of the Hunt Institute and Ruth Schallert of the Smithsonian Institution Libraries assisted in preparing the catalogue text. James White, Mary Kay Johnsen and Robert Kiger read drafts, made many useful suggestions, and helped to eradicate errors.

GB.

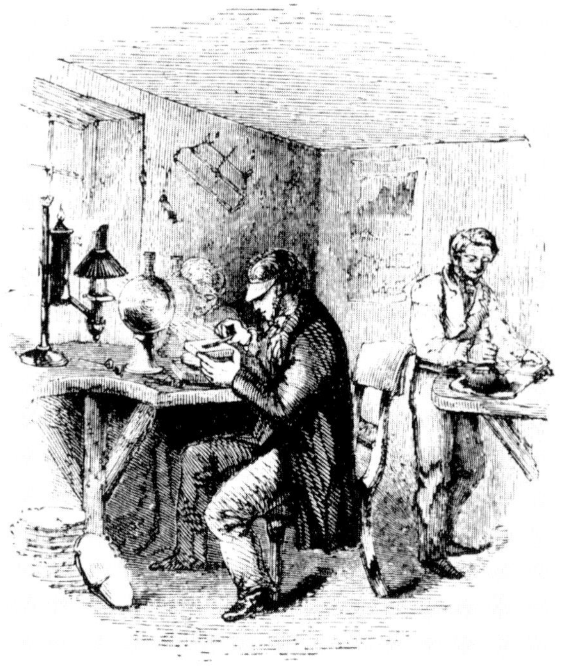

PRINTING SURFACES AND PRINT PAPER

Magnifications of some typical kinds of printing surfaces:

Figure A. *Relief: Woodcut.* A plank-grain block cut by Wolfgang Meyerpeck in 1563. The stoutness of the lines is quite apparent, despite a thick deposit of old congealed ink that coats all the interstices. This was the original block used for printing item #5.

Figure B. *Relief: Wood engraving.* A boxwood block engraved by Thomas Bewick in 1782. The extremely hard end-grain surface permitted a much finer quality of line than did the plank-grain surface of the early woodcut. Both black and white lines are clearly visible, the latter showing as fine graver cuts against a black background. This type of block was used for printing item #42.

Figure C. *Relief: Zinc etching.* A relief, photoengraved block reproduced from a pen-and-ink line drawing, reduced from the original and etched by the Durham Engraving Co. in 1968. This type of block was used for printing item #67.

Figure D. *Relief: Half-tone photoengraving.* A relief, photoengraved block reproduced from a photograph, made in 1903 (manufacturer unknown). The diagonal dot pattern created by the cross-ruled screen is vividly clear. This type of block was used for printing item #66.

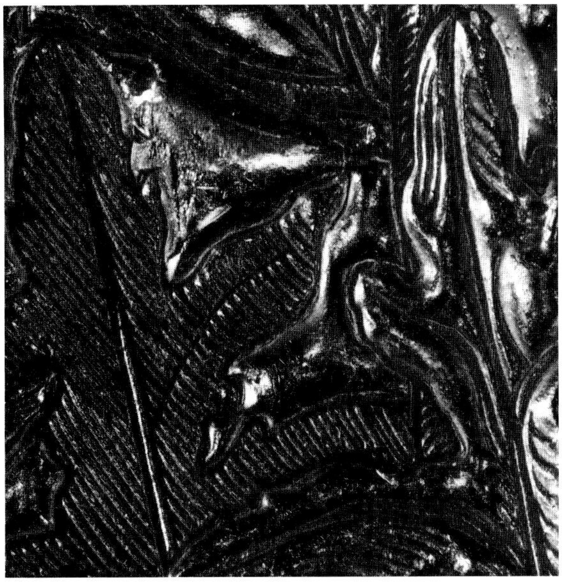

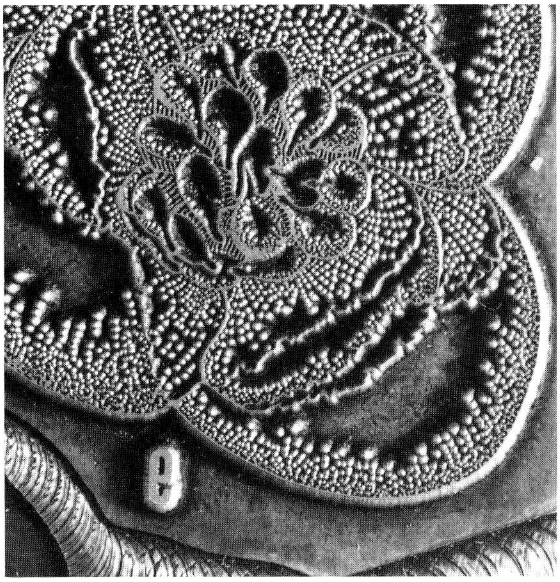

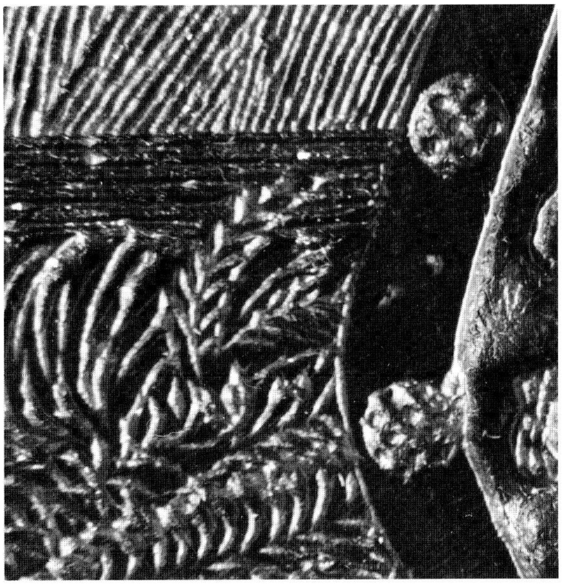

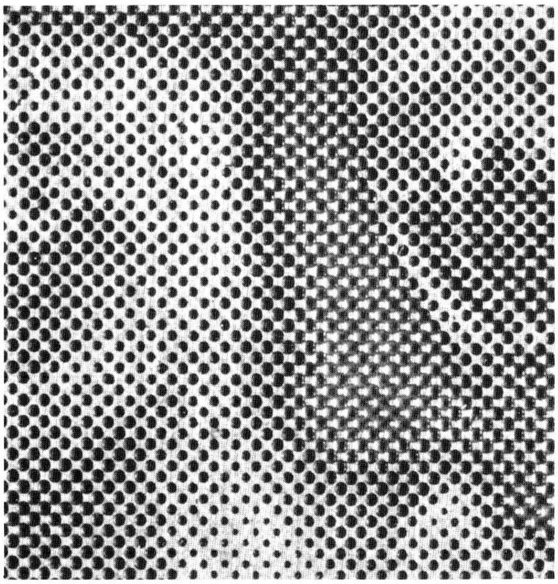

Figure E. *Intaglio: Line engraving.* A copper plate engraved and etched by John Curtis in 1812. The difference between the nature of the deep, cleanly cut burin lines and shallower, "bubbly" etched lines is clearly visible. This type of plate was used for printing item #16, for example.

Figure F. *Intaglio: Stipple Engraving.* A copper plate stipple etched by John Curtis in 1812. The dots are etched to varying depths, some looking as deep and round as potholes, others shallower and irregular in outline. This type of plate was used for printing item #25, for example.

Figure G. *Intaglio: Aquatint engraving.* A copper plate aquatinted by J. D. Smillie in the early 19th century. The etched reticulate-patterned tones are apparent. This type of plate was used for printing item #27.

Figure H. *Planographic: Lithograph.* A lithographic stone with a drawing made by Joseph Prestele around 1850. The image and the background are completely flat. This type of stone was used for printing item #53, for example.

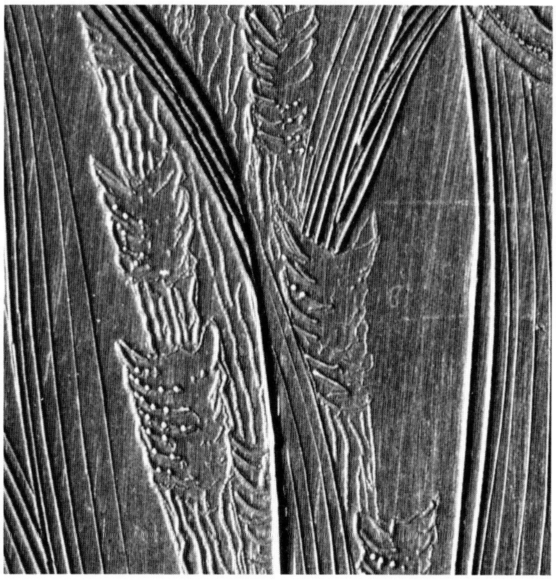

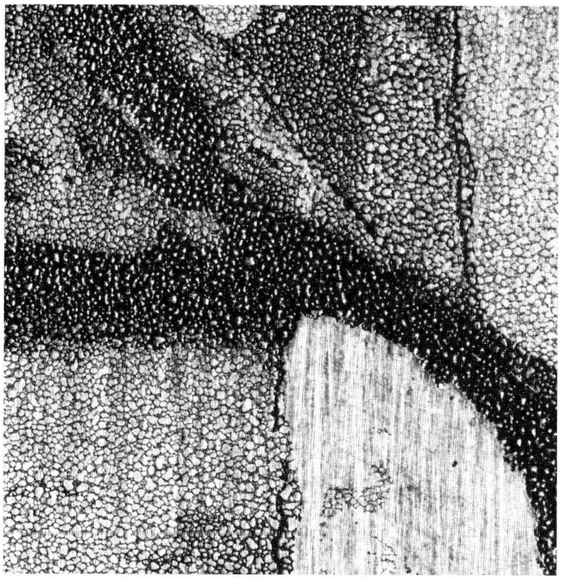

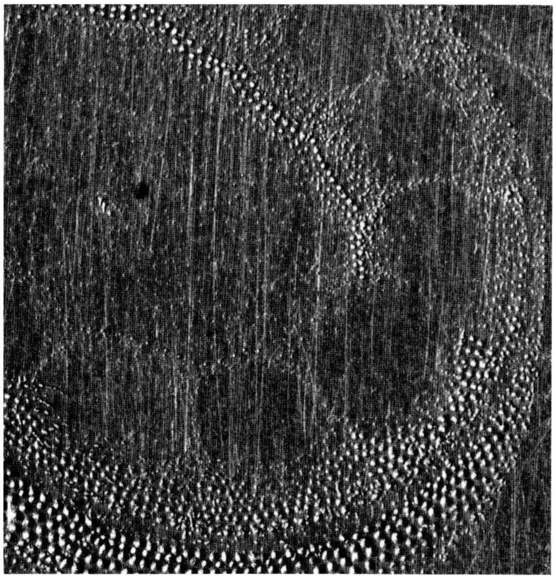

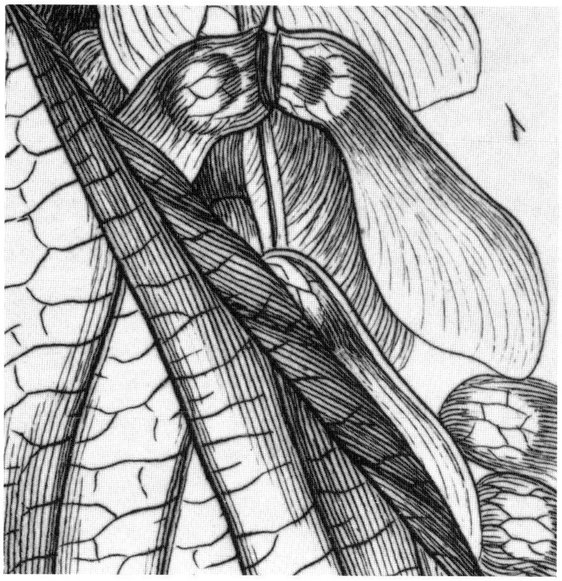

Magnifications of the surfaces of some typical kinds of paper:

Figure I. *Laid paper, 1485; from item #1.* The coarse surface texture includes the fiber impressions of the felt used in pressing and drying the sheet of paper during manufacture.

Figure J. *Laid paper (wire side), ca. 1753; from item #10.* The fine parallel grooves are impressions of the traditional papermaking mould. These "wire" lines are crossed, at right angles, by "chain" lines that usually lie about 1 inch apart.

Figure K. *Wove paper (wire side), 1826; from item #40.* In the later 18th century a new kind of mould was introduced to paper manufacture. It had a very fine woven wire mesh that left only a faint cloth-like impression.

Figure L. *Machine-made paper, 1895; from item #43.* The surface of typical machine-made book paper, from the mid-19th century onwards, was normally free of any discernible pattern. A specially smooth surface was required for delicate relief-printed illustrations.

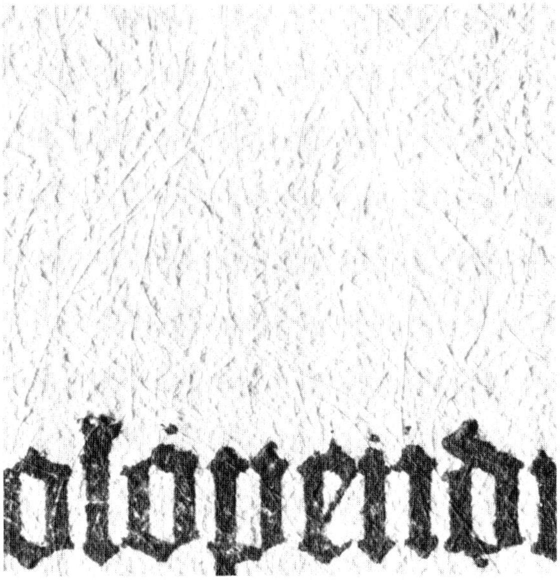

Figure M. *Paper surface altered by printing.* A "blind" printed etching showing embossed surface and platemark created by printing the plate without ink. Printed on moldmade paper, 1985.

Figure N. *Diagrammatic cross-section of impressions from relief, intaglio, and planographic printing surfaces.* Drawn with exaggerated contours to demonstrate the way that ink lies on the paper.

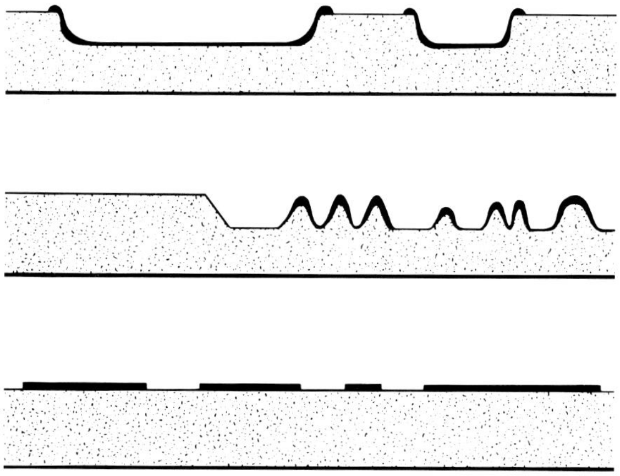

Catalogue

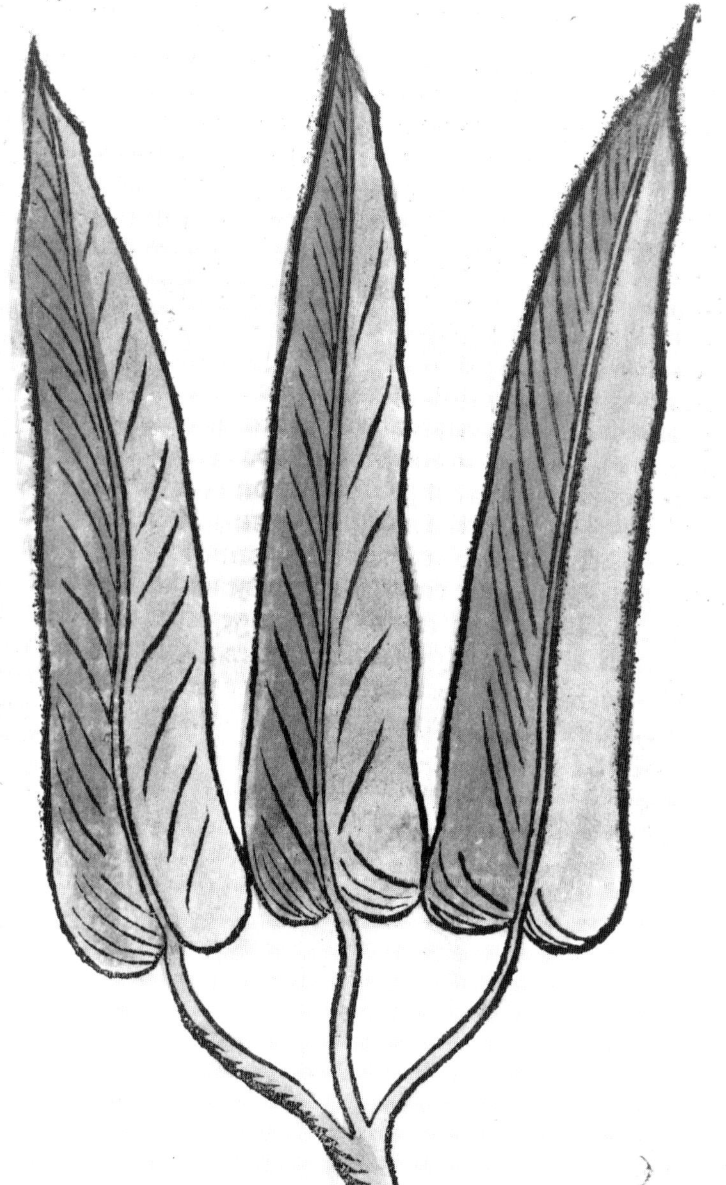

Scolopendria hirtz zunge Cap̄·ccclj·

Scolopendria latine· grece aplenon vel aplniū vel splemon vel scolopendriū·arabice ceterach· Diascorides in dē capitel aplenon id est scolopendria beschribt vns vñ spricht das diß bletter glichen an der gestalt der slangen scolopendria geheyßen vñ die slange hat·viiii·füß· Diß wechset gern an den felschen oder an den steynen muren·

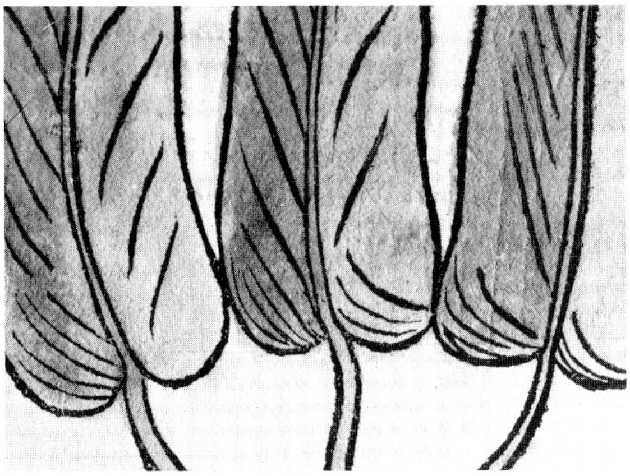

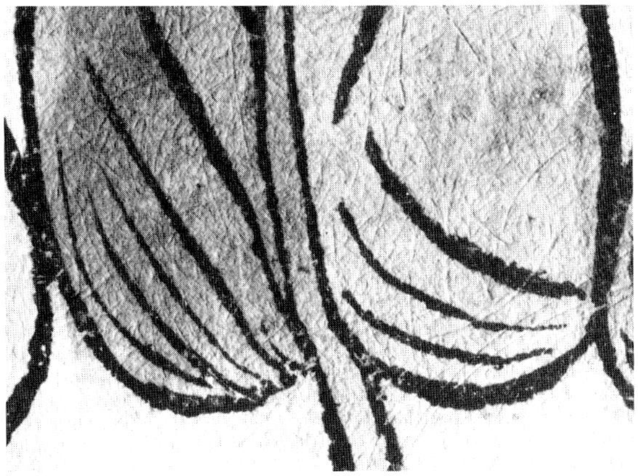

1. Scolopendria hirtz zunge. Coarse line woodcut with minimal line shading, artist and blockcutter unknown, with accompanying text. Forming an illustrated page (type area 20.6 x 12.3 cm.) printed at Mainz in 1485 by Peter Schöffer. Printed in black ink on laid paper (wire side, coarse felt impression, sheet size 26.6 x 20 cm.). Hand-finished with (later?) opaque watercolor. Ex: [Gart der Gesundheit. Mainz, P. Schöffer, 1485]. Cap. CCCLI. Nissen BBI #2266.

The earliest printed botanical pictures appeared in the mid-1480s in illustrated treatises on herbs and their uses. For a time when the return to naturalism in manuscript herbal illustration was already firmly established and when various woodcut images of delicacy, intricacy and charm were regularly being produced, it is sad to see such stylized, distorted and derivative figures in the printed herbals.

Woodcuts of this period were cut on "planks" of fairly soft fruit wood (i.e. planed with a lengthwise grain) using sharp, very small-bladed knives for line detail and gouges for clearing away large spaces between. The relief images so produced were inked with a mixture of boiled linseed oil and lampblack and printed on the hand press in the normal way. They could therefore be set in the same forme as any accompanying letterpress to create an illustrated page.

This rugged, stout-lined figure of a hart's-tongue fern, prepared by an unnamed block cutter, is from the so-called "German Herbarius." It is one of the 369 plant illustrations after various artists, most of which share its simple, bold delineation and distorted portrayal. The whole book was printed on a coarsely textured paper, and the woodcut lines have sunk deep into the surface and produced heavily impressed images. The thickness of the outline on this block was further coarsened by such careless overinking that accumulated deposits have printed as a ragged fuzz outside the woodcut line—very clumsy work to have come from Peter Schöffer's famous press. The woodcuts lack any meaningful shading or modelling since it was intended that they should be finished with hand coloring just as this one is (though in this case the coloring is almost certainly of a later date).

This essentially medical work, put together by a well-to-do amateur with the assistance of the physician Johann von Cube, described the virtues, natures and colors of herbs. Some 65 of the illustrations were apparently drawn from actual plants but the great majority share the uninformative simplicity of this example and could have had little more than decorative value for the reader.

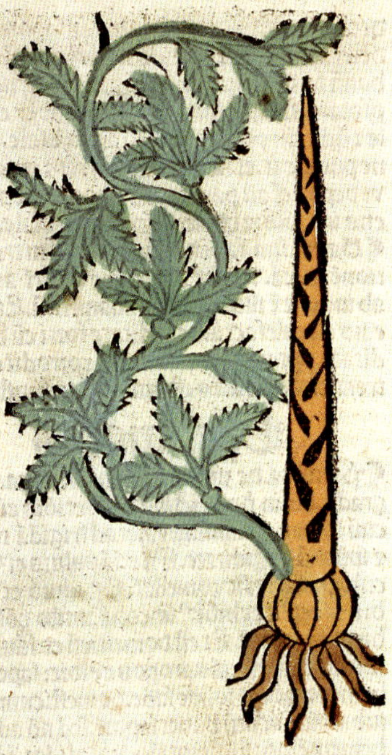

ASparagus siue asparagꝰ latine ꝛ gre-
ce. Ara. nalion vl' halion. Serap. li.
aggre. ca. nalion auc. Gal'. i. zparagꝰ.
virtꝰ eiꝰ ē q̄ desiccat ſ; nō h; frigiditatē; neq;
caliditatē. ꝛ abstergit ꝛ apit opilationē epat;
ꝛ renum. ꝛ proprie semen ꝛ radix.

Operationes

A ¶ Sera. Zparagus curat dolorē; dentiū. q̄ de-
B siccat sine calefactiōe ꝛ infrigidatōe. ¶ Et idē
auc. Dyas. Q̄ñ elixat' vna ebulitiōe ꝛ come-
C dit'. mollit ventrē ꝛ puocat vrinā. ¶ Et q̄ñ co-
quit' radix eiꝰ ꝛ bibit' decoctio ipsiꝰ ꝯfert strā-
D guirie ꝛ ycterície ꝛ sciatice. ¶ Et q̄ñ decoquit'
in vino ꝯfert decoctio eiꝰ morsui rutele.
E ¶ Et q̄ñ ex decoctiōe eiꝰ fit collutio oris cō-
fert dolori dentiū. ꝛ semen eius q̄ñ bibit' facit
F opatiōes radic;. ¶ Et dr̄ q̄ elixatura eiꝰ occi-
dit canes. ꝛ dicūt aliq̄ q̄ q̄ñ cornua arietū se-
G peliunt' in terra nascuntur inde sparagi. ¶ Et
idem auct. Aben mes. ꝓprietas est remouere
dolorē lumboꝝ q̄ fit a flegmate ꝛ ventosita-
te: ꝛ dolorē colice. q̄ mollit ventrē. ꝛ si hō vti-
H tur eo multū facit nauseā. ¶ Galienꝰ. ij. de ci-
bis ca. de Zparago ait: q̄ sunt diuretici mo-
dicū hñtes nutrimēti ꝛ nō bene digerunt'.
I ¶ Et idē. vi. sim. far. ca. zparagꝰ est abstersiue

virtutis: nō tñ calefacit manifeste: neq; infri-
gidat. Et iō tam renu; q̄z epatis deopilatiuꝰ
est: ꝛ mag; radices ꝛ semīa q̄z herba. q̄m etia;
sanat dolorē dentiū. quoniā sine calefactiōe
desiccat. quorum egent maxime dentes.

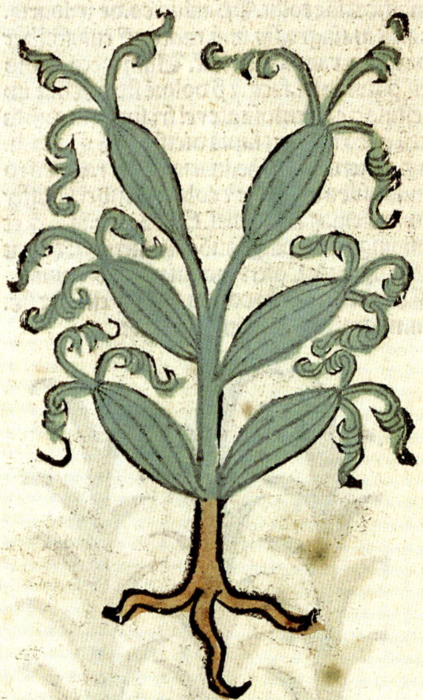

Capitulum. cccccxxiiij.

ZEduar. Auicenna. Zeduar me-
lior est illa que cum napello cre-
scit. et eius vicinitas plantam
napelli debilitat.

Operationes

¶ Est autē tyriaca venenoꝝ omniū viperi et A
napelli ꝛ alioru;. ¶ Rasi. Zirumbet: id est ze- B
duaria. ca. est ꝛ sic. ventositatē dissoluit.
¶ Constantinus in libro graduum. Zeduar C
ca. est in. iij. sic. in primo. valet contra venena
ꝛ morsus reptiliū. stomachum curat ꝛ confor-
tat. appetitum excitat. Putorem oris ampu-
tat: si post allia accipiatur.
¶ Platearius. Zeduar ca. est in. iij. gra. sic. in D
ij. Vinū decoctiōis eius ꝯtra frigidā tussim
valet. ꝛ ꝯ dolorē stōachi ꝛ intestinoꝝ ex frigi-
ditate ꝛ ventositate. ¶ Salsamentū ex eo fa-
ctū ꝛ rore marino ꝛ pane asso ꝛ aceto appeti-
tum excitat. Suppositorium ex eo et trifera
matricem calefacit ꝛ mundificat.

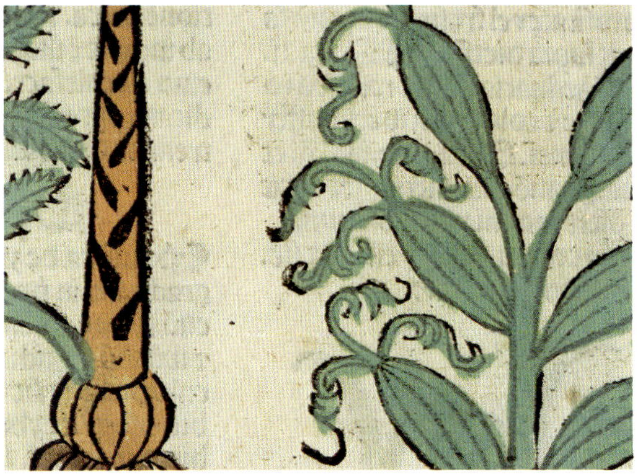

2. Zparagus. *and* Zeduar. Coarse line woodcuts, artist and blockcutter unknown, with accompanying text. Forming an illustrated page (type area 23.6 x 15.8 cm.) printed at Strasbourg ca.1499 by Johan Prüss. Printed in black ink (image areas ca. 10.5 x 6 cm.) on laid paper (wire side, sheet size 29.8 x 20.8 cm.). Hand-finished with opaque and transparent watercolor. Ex: *Ortus Sanitatis. De Herbis et Plantis...* [Strasbourg, J. Prüss, ca.1499]. Leaf li iiii recto. Nissen BBI #2365.

The 1490s, which saw the production of some of Dürer's early woodcut masterpieces, also bequeathed this "garden of health," a substantial tome with herbal woodcuts of childishly awkward simplicity, mostly derived from the 1485 "German Herbarius" already seen. It was first published in 1491 at Mainz, and variously republished; this is Prüss' Strasbourg reprint of ca.1499, with newly prepared woodcuts.

Although the text of this work is much fuller than that of the "Herbarius," its 530 botanical illustrations display a return to a more mediaeval character than the images they imitate, and they reveal a generally faulty understanding of the natures of the plants portrayed. They are smaller in size, even less carefully cut, less informative, and are dropped into double-column page settings with scant regard for harmony with the typographic design. All are finished with heavy, stylized coloring—almost certainly contemporary in this instance. Once again the printing is "ragged," much more so than is apparent for the text. Possibly the motion required for inking a bed of type was really unsuitable for the widely spaced lines of woodcuts, but since both occupied the same forme, it was not possible to give locally different inking. Dürer's famous refined and intricate woodcuts, in comparison, were obviously awarded much greater care in both inking and printing.

The extensive republication of this work must be accounted a measure of its textual value for, as with the previous exhibit, these caricature portrayals can scarcely have satisfied the botanical interest of any reader.

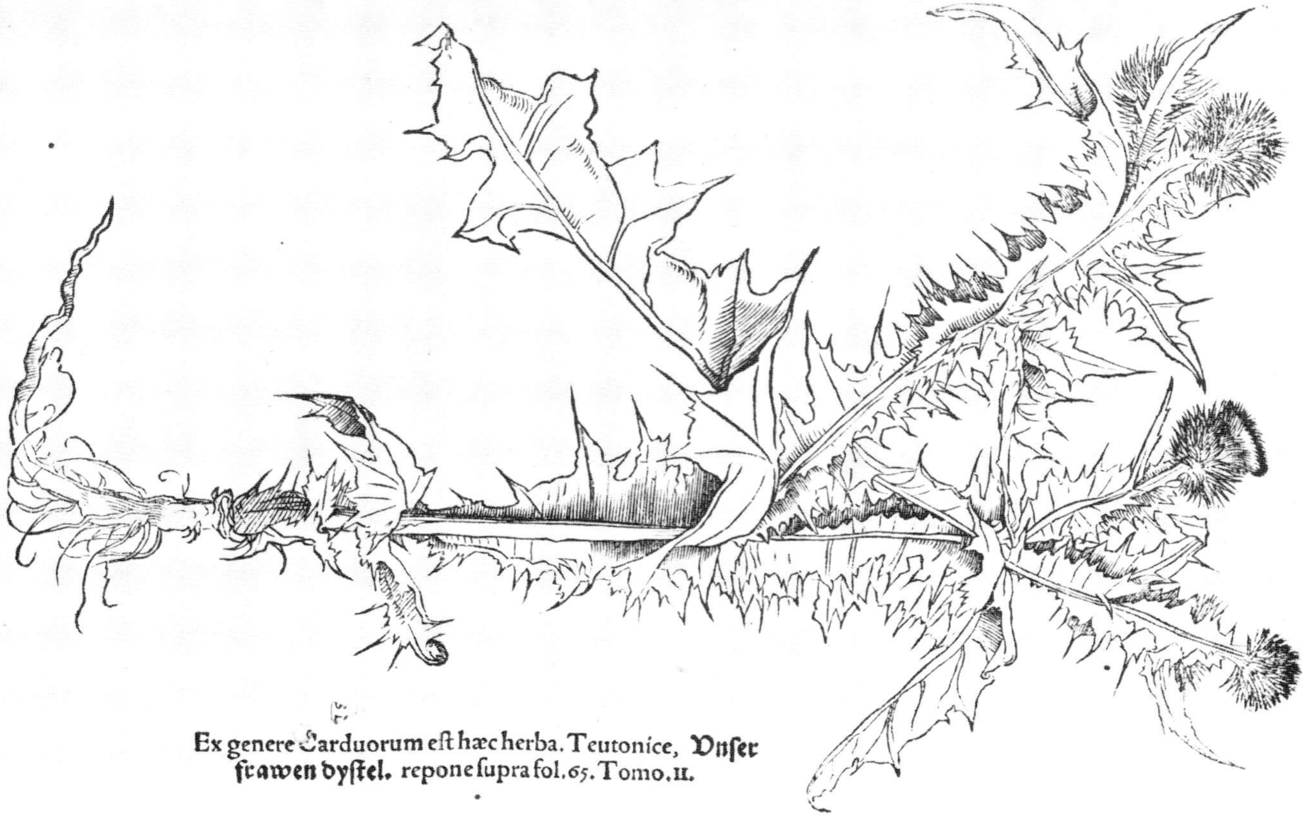

Ex genere Carduorum est hæc herba. Teutonice, **Unser frawen dystel.** repone supra fol. 65. Tomo. II.

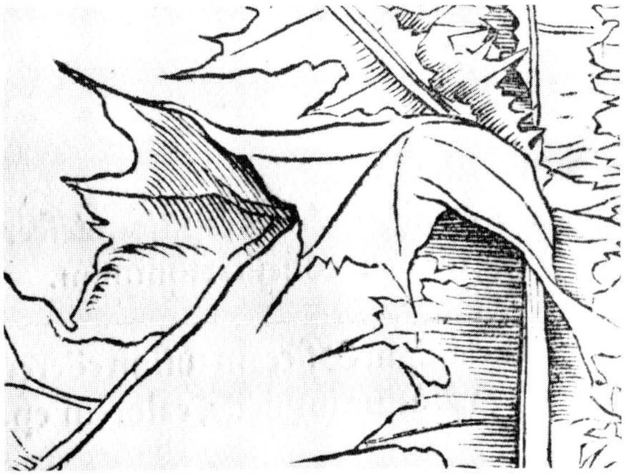

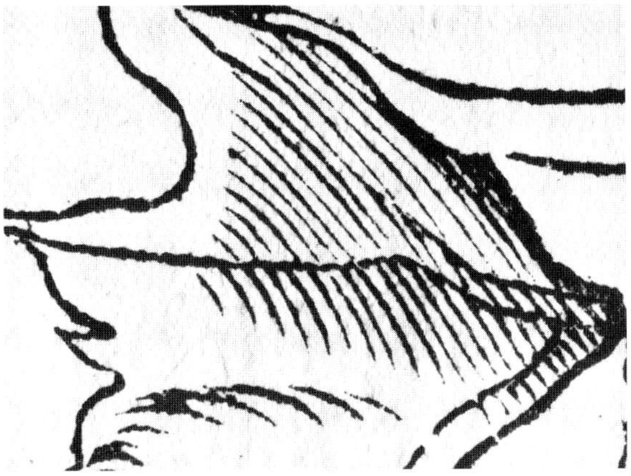

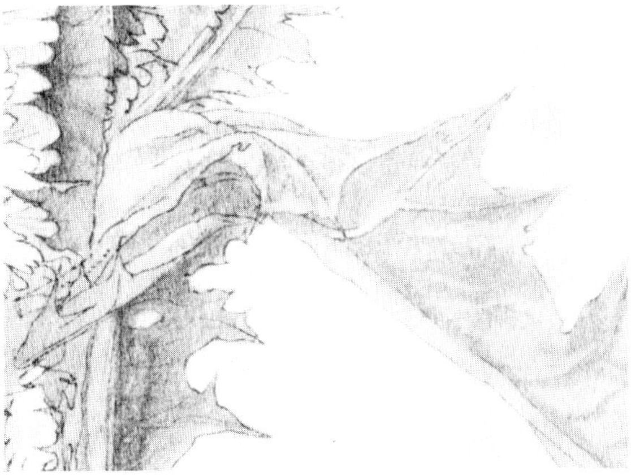

3. Ex genere Carduorum est haec herba. Teutonice. Duser / frawen bystel. Line woodcut with moderate use of line shading, blockcutter unknown, after a drawing by Hans Weiditz (German ca.1500-1536). Forming a full-page illustration printed at Strasbourg in 1530 by J. Schott. Printed in black ink (image area 25.6 x 16.6 cm.) on laid paper (wire side, page size 31.4 x 20.4 cm.). In: O. Brunfels. *Herbarum Vivae Eicones. (Tomus Herbarii Othonis Brunfelsii III).* Strasbourg, J. Schott, 1536. Folio 65. Nissen BBI #257. Shown together with color process-engraved facsimile of the original watercolor drawing, in: W. Rytz, ed. *Pflanzenaquarelle des Hans Weiditz aus dem Jahre 1529.* Bern, 1936. Plate 15.

Given this work boasting the title "Living portraits of plants," the 16th-century botanical reader could have expected to find pictures that honestly portrayed the characters of plants as he saw them in the field. A glance at its contents would have reassured him, for here was a break from the conventionalism of earlier plant illustrations.

Brunfels' work was illustrated by Hans Weiditz, a gifted and observant artist who consistently drew from nature with great fidelity. Indeed, the uncompromising "warts and all" honesty of his plant portraits gave rise to some criticism of his work. Most of the woodcuts are large, occupying full pages, but some others are no more than an inch or two high.

We are able to judge the quality of both the artist's draughtsmanship and the unnamed blockcutter's skill because some 77 of Weiditz's originals survive to this day in the University of Berne. We show a modern facsimile for comparison with the woodcut. The artist's bistre outline formed a pattern for the blockcutter's outline, whilst the coloring suggested the modelling that he had to translate into line shading. It is immediately apparent that he made certain alterations in fitting the whole plant into the page without reducing the size of the flowers and leaves. Even so, his image is somewhat larger than the type area of other normal letterpress pages. In general the blockcutter worked with a confident, free, and clean line, kept his shading to a necessary minimum, and made a decent translation of Weiditz's original into the black and white language of the 16th-century woodcut.

Linnaeus found these illustrations acceptable enough to constitute part of the historical basis for certain species described in his great Species Plantarum *of 1753.*

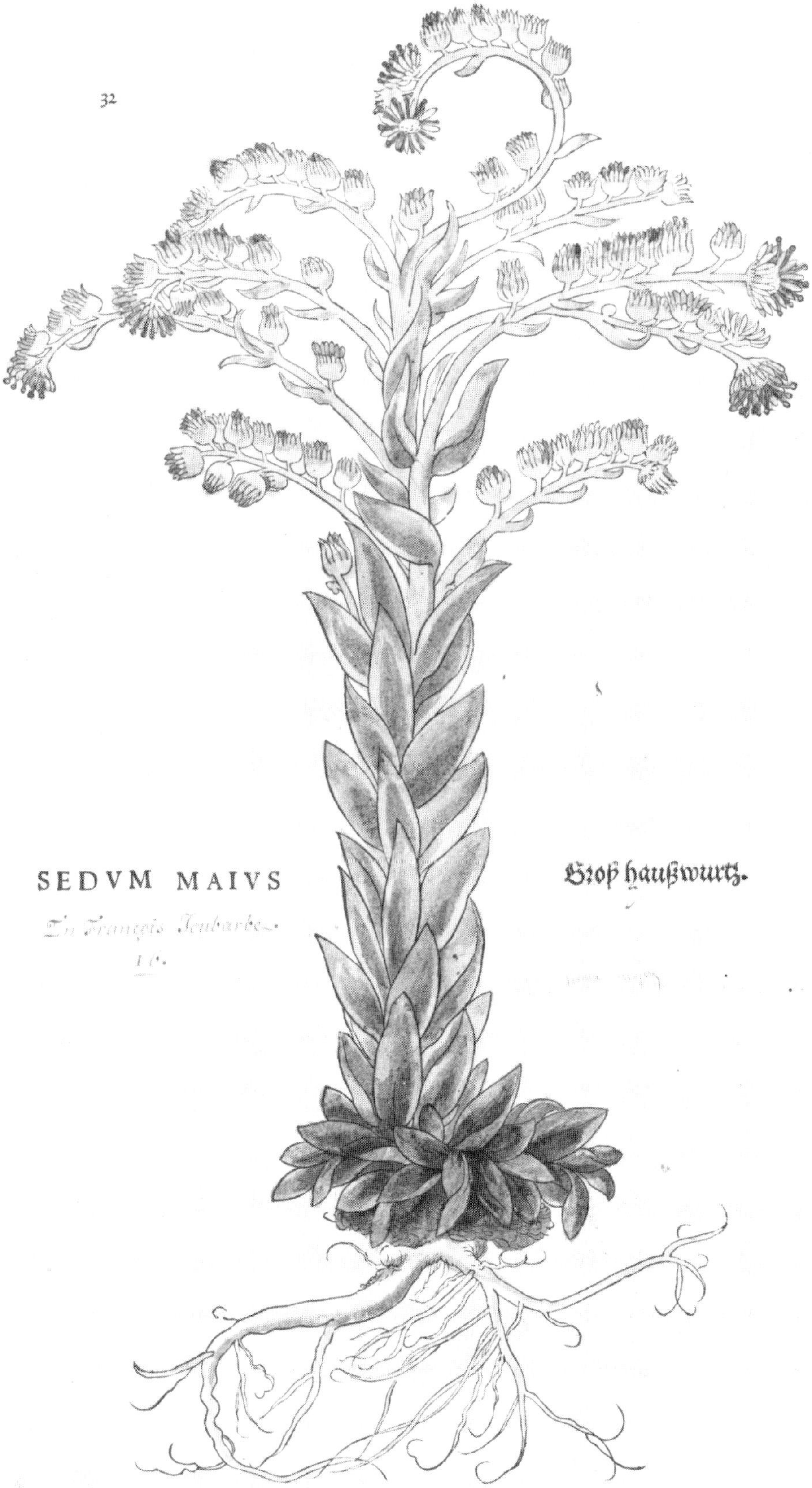

SEDVM MAIVS

En Françoys Joubarbe.

Groß haußwurtz.

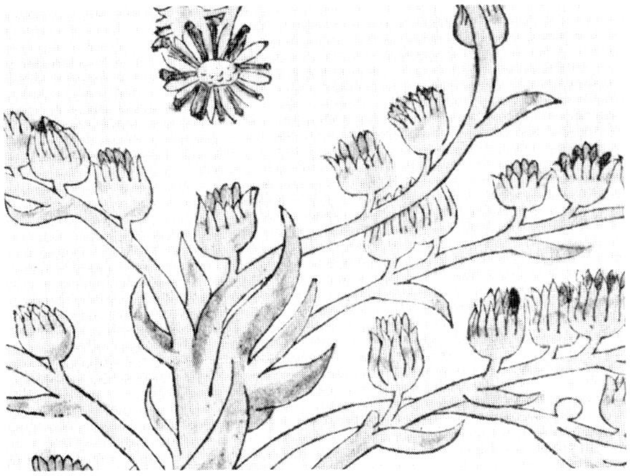

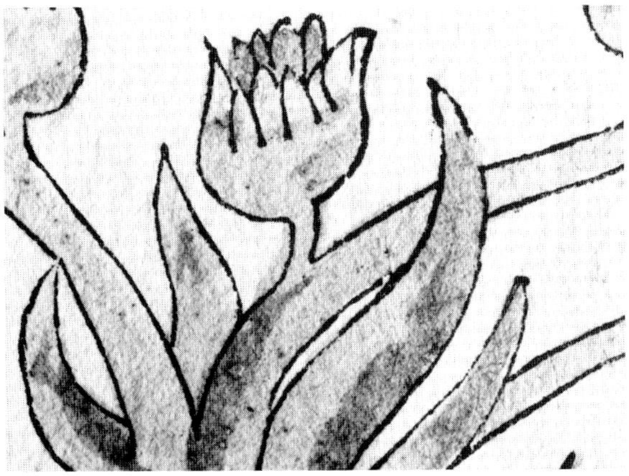

4. SEDVM MAIVS Gross hausswurtz. Fine outline woodcut, engraved by Veit R. Speckle (German? fl. 1540s) from a drawing by Albrecht Meyer (German fl. 1540s), redrawn on the block by Heinrich Füllmaurer (Swiss fl. 1540s). Forming a full-page illustration printed at Basle in 1542 by M. Isingrin. Printed in black ink (image area 31.8 x 18 cm.) on laid paper (wire side, sheet size 35.2 x 23.3 cm.). Hand-finished with (later?) transparent watercolor. Ex: L. Fuchs. *De Historia Stirpium.* Basel, Isingrin, 1542. Page 32. Nissen BBI #658.

The year 1542, some 57 years after the appearance of our primitive first exhibit, saw the publication of an illustrated botanical work that forms the high-water mark of the early botanical woodcut tradition. Fuchs' De Historia Stirpium *was the product of his refined taste, the skill of his artists, and much expenditure, and is generally recognized as the starting point of floristic publication. Its crisp white folio pages provided space for 519 large woodcuts, life-size where possible, and drawn directly from nature. They included about 400 native German plants, the first ever attempt to cover the flora of a locality.*

Meyer's original drawings are preserved at the Österreichische Nationalbibliothek in Vienna. Their pen-drawn outlines and watercolor finishes were admirably reproduced in the published woodcuts. Meyer's designs were first copied onto the block by Füllmaurer, and Speckle, "the best engraver in Strasbourg," cut the fine outline relief images. His woodcut line, which A. H. Church estimated was about 250 micrometers in width, could not permit the depiction of such fine details as hairs, stamens or parts of florets of less than 1-2 mm. diameter. However, within this limitation the printed images conveyed valuable botanical information. Whole plants, flowers, stem, leaves and roots were all shown, and only when page dimensions dictated, was any "rectangularization" of plant habit allowed. There is however some idealization of depicted plant specimens, and flowering and fruiting are sometimes shown on the same plant.

All the images had plenty of enclosed unshaded white space, since the publisher intended that they be colored. Well-finished colored copies were issued by the publisher, and it is believed that Meyer may have supervised their completion. Certainly, authentically colored copies appear to follow the originals with great fidelity. Plain copies were also published and some were subsequently colored by their owners—which is probably true of the exhibited example. Since the figures were drawn and cut in only bare, if elegant, outline, it was the coloring that provided the necessary modelling and shading as well as other botanical information.

ERVCA SYLVESTRIS.

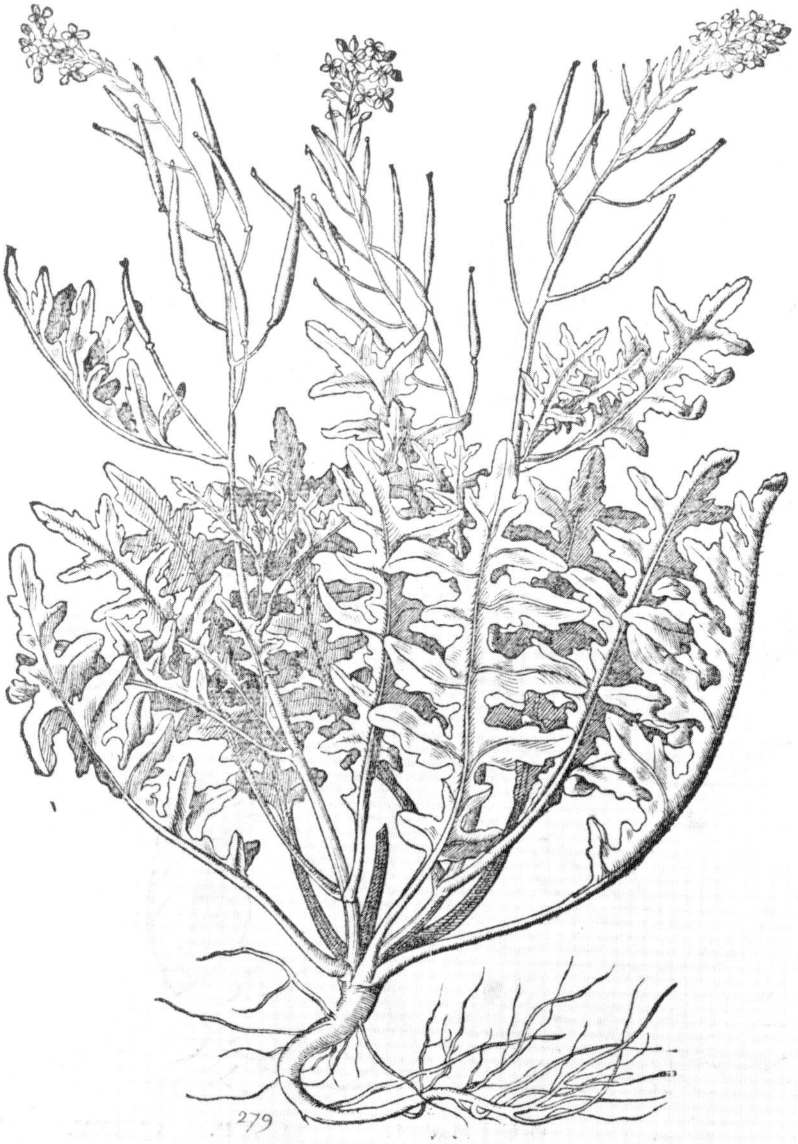

par frigus admiſtum exceſſum temperat. Quinetiam genitale ſemen augere creditur, & in uenerem ſtimulos addere. Caput etiam magis dolore afficiet, ſi ſolum edatur. Semen (ut ſcribunt antiquorum nonnulli) muris aranei morſibus medetur. Ventris tineas enecat, lienemq́; extenuat. Tritum, & cum felle bubulo miſtum, nigras ulcerum cicatrices ad candorem reducit. Cum melle uerò perunctum, uitia cutis in facie emendat, & lentigines emaculat. Herba, quæ Græcis εὔζωμον, Latinis Eruca uocatur: Mauritanis, Iergir, Ergit, ſeu Giargir: Italis, Ruchetta, & Rucola: Germanis, Vueiſz ſenff: Hiſpanis, Oruga, & Arugua: Gallis, Roquette. Nomina.

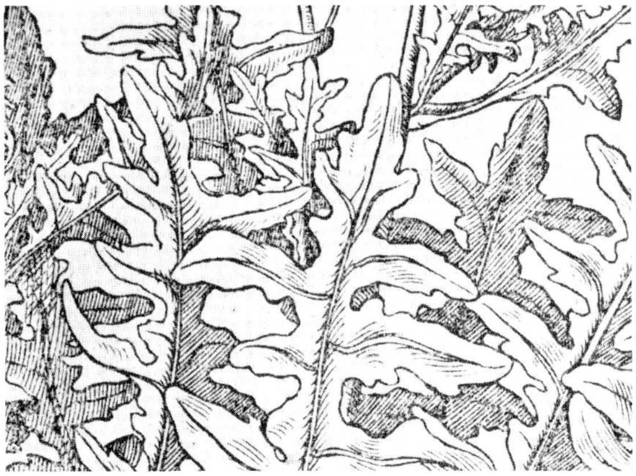

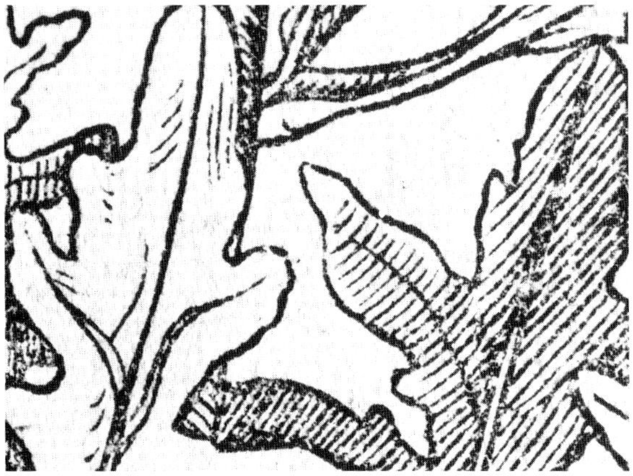

5. ERVCA SYLVESTRIS. Line woodcut with some cross-hatched shading, engraved by Wolfgang Meyerpeck (German fl. 1560s) from a design probably drawn by Giorgio Liberale (Italian 16th cent.), with accompanying text. Forming an illustrated page (type area 30.1 x 18.7 cm.) printed at Venice in 1565 by Felice Valgrisi. Printed in black ink (image area 22 x 16 cm.) on laid paper (wire side, sheet size 34.5 x 21.8 cm.). Ex: P. A. Mattioli. *Commentarii in Sex Libros Pedacii Dioscoridis Anazarbei de Medica Materia.* Venice, Ex Officina Valgrisiana, 1565. Page 531. Nissen BBI #1305. Shown together with impressions of two later, smaller imitations, ex: J. Parkinson. *Theatrum Botanicum.* London, T. Cotes, 1640. Page 818. Nissen BBI #1490; and M. de L'Obel. *Plantarum seu Stirpium Icones.* Antwerp, C. Plantin, 1581. Page 204. Nissen BBI #1220

Matthioli's expansive commentary on Dioscorides' classic text, illustrated with over 500 figures of plants, is reputed to have sold upwards of 30,000 copies in its early editions. The sheer bulk of this tome reflects the growing accumulation of botanical knowledge, which is curiously mirrored also in its illustrations. They are generally presented in woodcuts of considerable size and unprecedented complexity, though some take their origin from earlier published sources such as Brunfels and Fuchs.

The designs, probably made by G. Liberale of Udine, were translated by Meyerpeck's accomplished block cutting into morphologically detailed and carefully shaded images whose style contrasts notably with the airy, simple elegance of Fuchs' illustrations. Apart from the close massing of foliage, fruit and flowers, such details as veins and even hairs are often depicted or suggested with great skill. The limiting factor was the achievable fineness of a woodcut line sturdy enough to withstand the heavy pressure of the printing press, a limit that had already been reached by Speckle in cutting the blocks for Fuchs. Whereas Speckle's blocks relied on the colorist to add informative detail, Matthioli's engraver attempted to convey all possible information in black line images that were not intended to be colored. Yet, while these works attempted botanical perfection within the medium, Blunt's view is that "for all their skillfulness...[they] mark the beginning of the decline of wood engraving [sic]."

The example shown contains a modest amount of detail compared with some of Matthioli's grander illustrations, but carries the peculiar interest that it may be compared with the very block from which it was printed over 420 years ago. If some of Matthioli's illustrations were derived from earlier sources, his in turn suffered the same attention by others, as the accompanying imitations dating from 1581 and 1640 demonstrate.

The laborious skill needed to produce complicated blocks of this nature can be appreciated when one considers that 24 separate knife cuts were required to delineate the intersection of two pairs of parallel lines. The block had to be turned and twisted again and again as hundreds of knife cuts carved out the image. Should we wonder then if the engraver's line sometimes lacks vibrancy?

44

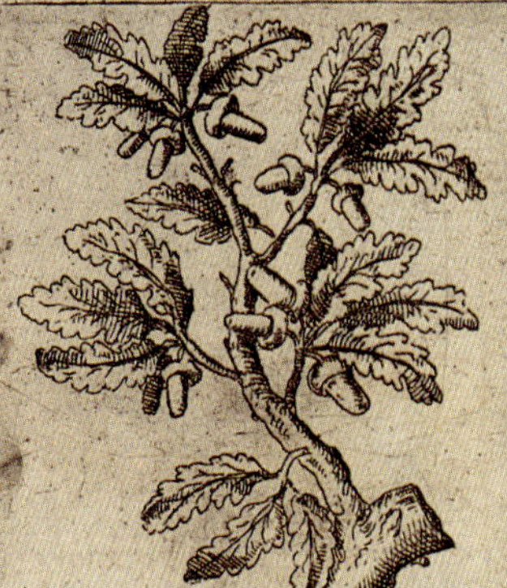

QVERCIA.

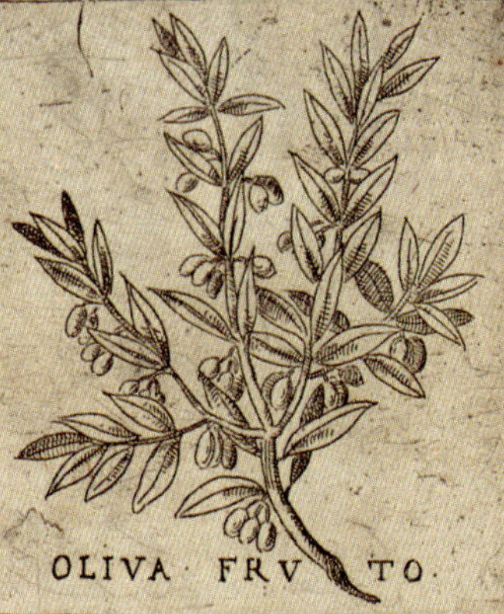

OLIVA FRV TO.

La decottione delle frondi uale a gli flussi desinterici, et stomacali et alo sputo del sangue, le giande prouocano l'orina, et mangiate generano dolor di testa –

Le oliue domestiche salate et condite peste et applicate non lasciano leuar le uesciche delle cotture del fuoco, et lauando la bocca con la salamoia ferma i denti

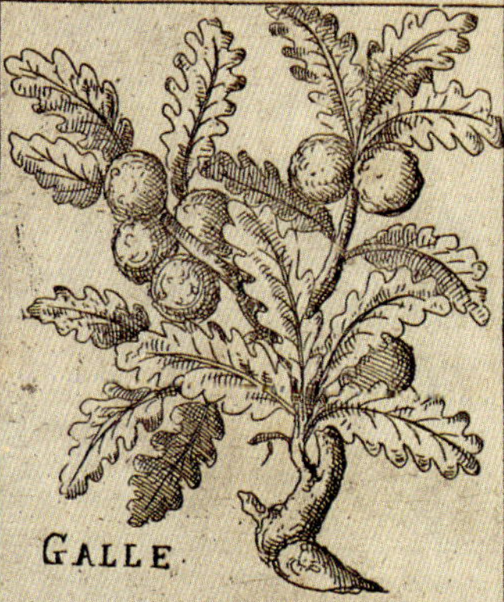

GALLE.

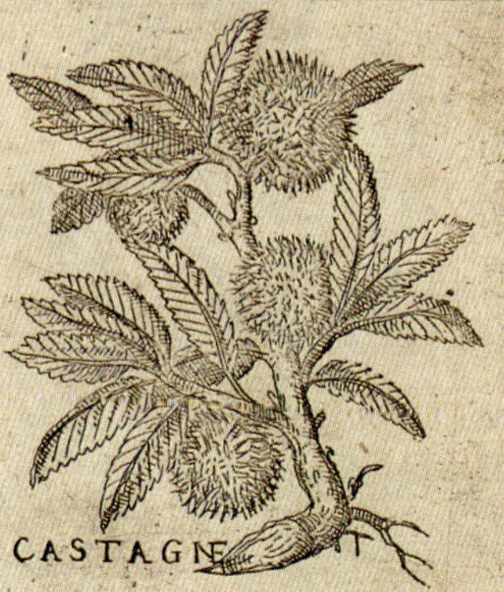

CASTAGNE

Trite in poluere rissoluono le su: perfluita della carne, ristagnano i flussi delle gengiue et del ugola, et saldano le uulcere della bocca et la carne cresciuta intorno alle gingiue –

23

pl. 23

Peste con mele et sale applicate uale al morso de li cani rabbiosi rissoluono le durezze delle mamelle. mangiate prouocano il coito, et fano doler la testa et generano uentosita –

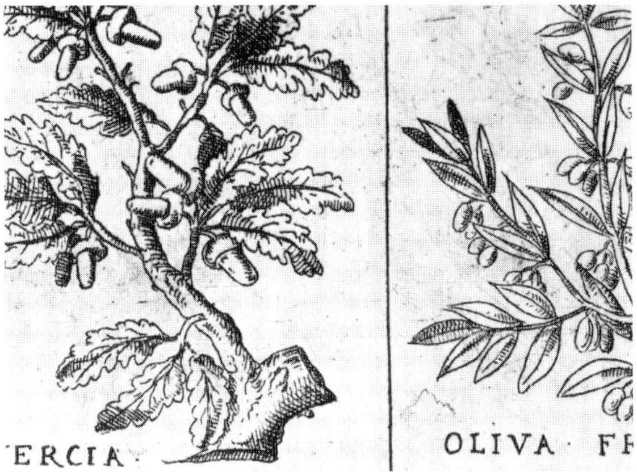

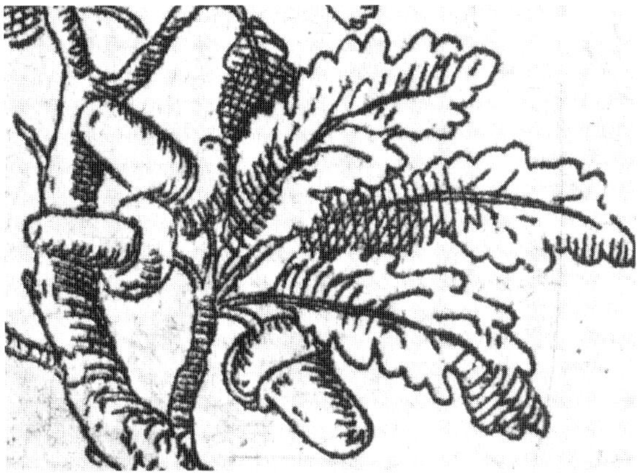

6. QVERCIA...OLIVA FRVTO... / GALLE...CASTAGNE ... Line etching with cross-hatched shading, single bite, with accompanying hand-written etched text, artist and etcher unknown. Forming an illustrated page printed at Rome ca.1580 by Pietro di Nobili (Italian fl. 1549-1584). Printed in brownish-black ink with distinct ink tone (platemark 18.5 x 11.5 cm.) on laid paper (wire side, watermarked "AB" astride an anchor, all within a circle, page size 24.7 x 19 cm.). In: P. di Nobili. [*Erbario Che in 32 Tavole Contiene la Figura di 128 Piante con la Dichiarazione delle Virtu e Proprieta di Ciascuna.* Rome], P. di Nobili, [ca. 1580?]. Page 23.

The intaglio printmaking method of incising or etching an image into a metal plate, rubbing ink into the lines, and printing from this surface, dates back well into the 15th century. The "Master of the Playing Cards" produced attractive engravings of birds, animals and a few plants as early as ca.1450, but more than a century had to pass before intaglio printmaking found its first botanical application.

The artist and etcher of these quaint figures are not known. Several of the pages bear the imprint "Petri de Nobilibus Formis" and on that evidence the work is dated as ca.1580. Nobili, a Rome publisher of prints and maps from ca.1545-1584, apparently prepared this unique little herbal in which text and illustrations are etched on the same plates. Only three copies of the book have been traced so far, and since they all lack a printed title page, it is possible that this work was never formally published. The watermarks in this copy suggest that it is a later printing done in the 1630s. Whatever its history, this earliest example of etched plant illustration opened a new chapter in botanical printmaking that was only partly closed when photomechanical printing arrived on the scene some three centuries later.

These deftly drawn figures have an autographic quality which suggests that the etcher was also the artist. Their simply sketched outlines and plentiful freehand shading provide easily recognizable thumbnail portraits of their subjects and, though somewhat stylized, they have fairly good modelling and perspective. The brief texts were written on the plate in mirror writing and each plate was apparently etched with a single "bite" (save, perhaps, the lettered names, which have a bolder line). A brownish ink was used for printing the finished page and the contrasting whiteness of the paper was muted by leaving a distinct and uniform "ink tone" over the whole plate.

The vernacular text has a directness which suggests that the work was intended as a pocket herbal vade mecum—*"QUERCIA The decayed leaves are to be used for flatulence and if there is blood in the sputum. The fruit is a diuretic and gives headache"; "CASTAGNE Make into a paste with apples and salt and apply to the bite of a rabid dog. Eating the fruit gives headaches and shortness of breath." The informality of the illustrations harmonizes nicely with the text.*

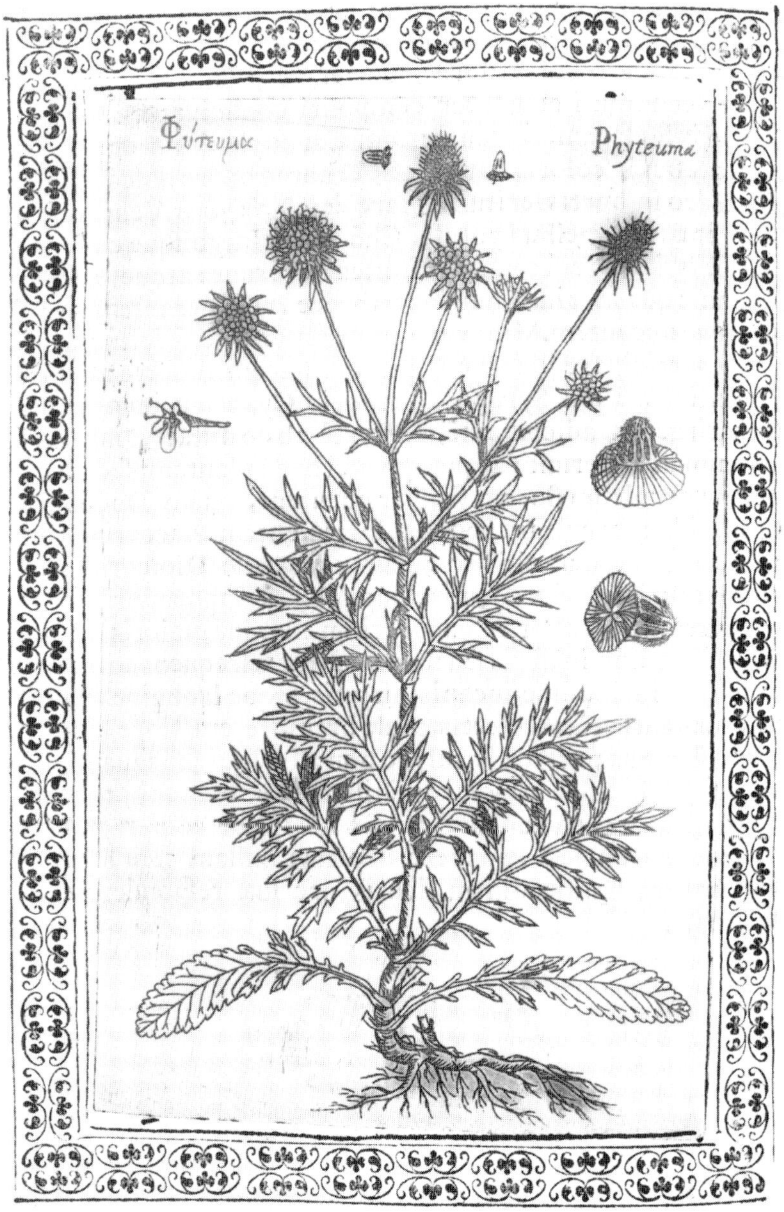
Φύτευμα
Phyteuma

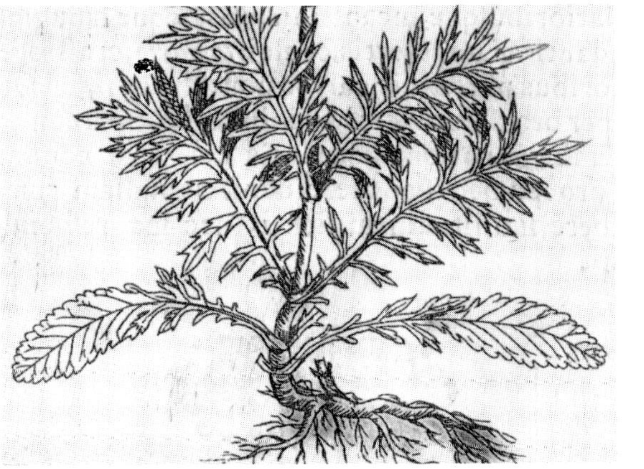

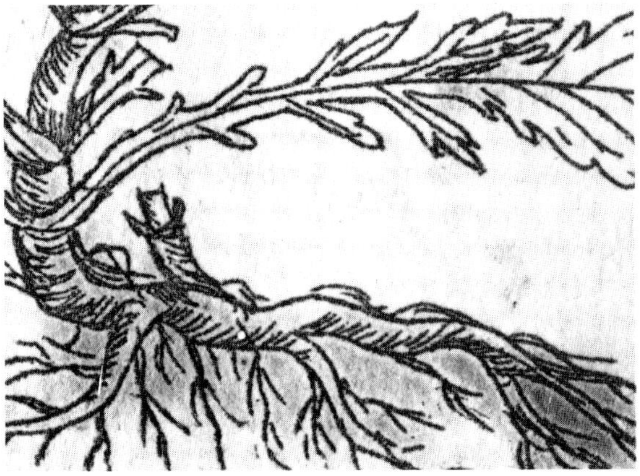

7. Phyteuma. Line etching, single bite, engraved by Fabio Colonna (Italian 1567-1640) after his own drawing. Forming a full-page illustration printed at Naples in 1592, plate printer unknown. Printed in umber ink with a heavy ink tone (platemark 13 x 8.2 cm.) within a typographic border of printer's ornaments printed in black (type area 16 x 10.2 cm.) on laid paper (top side, page size 20 x 14 cm.). In: F. Colonna. [*Phytobasanos,*] *sive Plantarum Aliquot Historia...* Naples, J. J. Carlinus & A. Paces, 1592. Page 98. Nissen BBI #386.

The introduction of intaglio methods for botanical printmaking offered some valuable advantages over the woodcut. It became possible for the artist to draw every line of the illustration with an etching needle exactly as he wished it to print, without the necessity of another hand to cut the printing surface. In botanical terms, autographic portrayal of plants by the author obviously implied greater understanding and authority.The slender etching needle gave the botanical artist the option of delineating fine details and thus, with relative ease, of illustrating more minute structures and textures than ever could have been attempted by a woodblock cutter. If he wished, he might leave the actual biting of his lines to an expert in the use of mordants.

It is believed that Colonna himself both drew and etched the 26 accurate illustrations to these commentaries on Dioscoridean plants, the original drawings for which survive in the Biblioteca Nazionale in Naples. Within limited space he provided a detailed line drawing of Phyteuma, roots and all, and added detailed figures of the minute flowers and seeds. The plate was given a single acid bite and printed in an unusual brown ink with a pronounced ink tone. It appears that the printer may have used local ink toning to accentuate some details of the etching.

One disadvantage of intaglio printing over the earlier woodcut was that it required a quite different printing operation from the relief letterpress. The mixing of intaglio and letterpress, as was done here, dictated that the sheets be printed twice. In this instance the printer obviously had some difficulty in aligning the etching within the typographic border. The overall result, though, is quite unusual and, to our eyes, attractive.

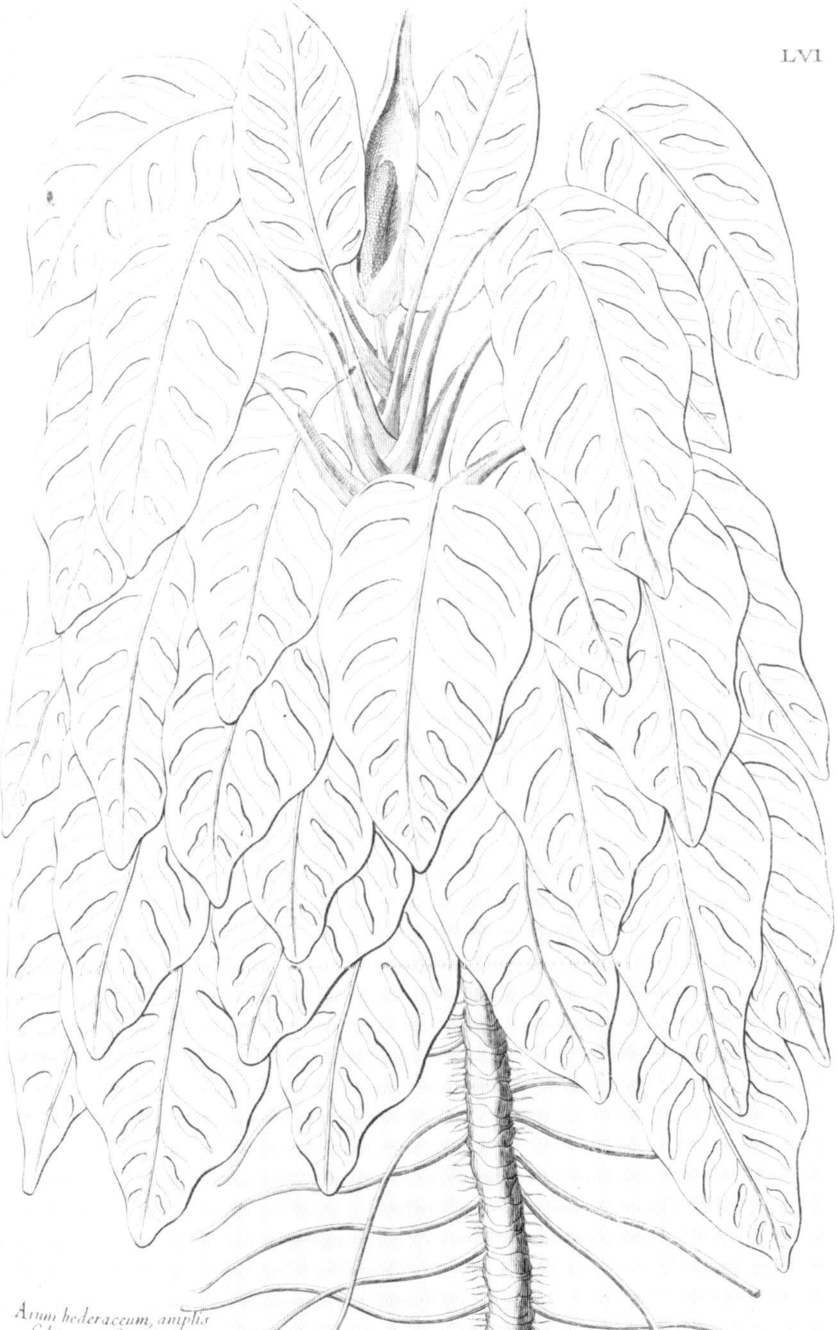

Arum hederaceum, amplis foliis perforatis.

Fr. Carolus Plumier Minimus Botanic. Regius delin.

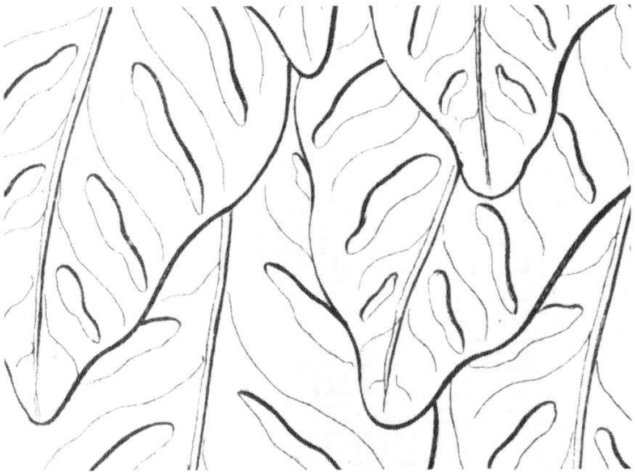

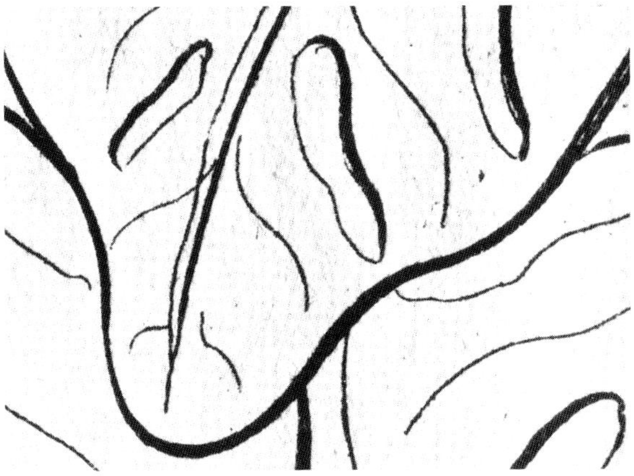

8. Arum hederaceum, amplis / foliis per foratis. Line etching, outline style with locally swelled lines, minor stippled and cross-hatched shading, single bite, etcher unknown, after a drawing by Charles Plumier (French 1646-1704). Forming a full-page illustration printed at Paris in 1693 by the Imprimerie Royale. Printed in black ink (platemark 37.8 x 24.5 cm.) on laid paper (wire side, page size 42.3 x 26.8 cm.). In: C. Plumier. *Description des Plantes de l'Amérique.* Paris, Imprimerie Royale [Jean Anisson], 1693. Plate 56. Nissen BBI #1544.

If the fineness of the needle-drawn line was a positive advantage in allowing detailed delineation of small features, it could prove a defect in larger-scale compositions. However springy and confident the outline of a large flower or leaf, the line was still apt to be thread-fine and lacking much force.

Plumier's grand outlines of West Indian plants would have been deprived of much of their vigor if the unnamed etcher, in copying his drawings onto copper, had not used a special kind of stylus to give a calligraphic strength to his line. He almost certainly used an échoppe, a much sturdier, untapered, needle. Instead of being sharpened to a point like a normal etching needle, it was cleanly sliced off at an acute angle so as to create an oval face, the distant curve of which gave the tip of the needle a tiny knife edge. When drawn edgewise through the soft waxy etching "ground" it would produce a needle-fine line, but if rotated so that the flat oval faced in the direction of travel, it would plough sufficient wax to create a broad line. With deft manipulation lines could be made to swell gradually from delicate fineness to sturdy boldness—and back to a hair-line again.

In this portrait of Arum *the effect was to add a measure of perspective and form to what was basically a very simple outline drawing. Dramatic emphasis was added by printing the plate in strong black ink on very white paper, the printer having cleanly wiped away all trace of ink tone. An alternative method of adding breadth to an etched line would have been to re-etch parts of it. The parts which were sufficiently etched would have to be "stopped out" with an acid-resisting ground before the plate was rebitten. For botanical work of this kind, however, the result would probably have had an undesirably mechanical character, since line strengthenings made in this way increased in abrupt steps.*

MUSA; FRUCTU CUCUMERINO LONGIORE,
quae in hypocausto D.ni Josephi Ayloffe Baronetti flores suos et fructus ferebat.

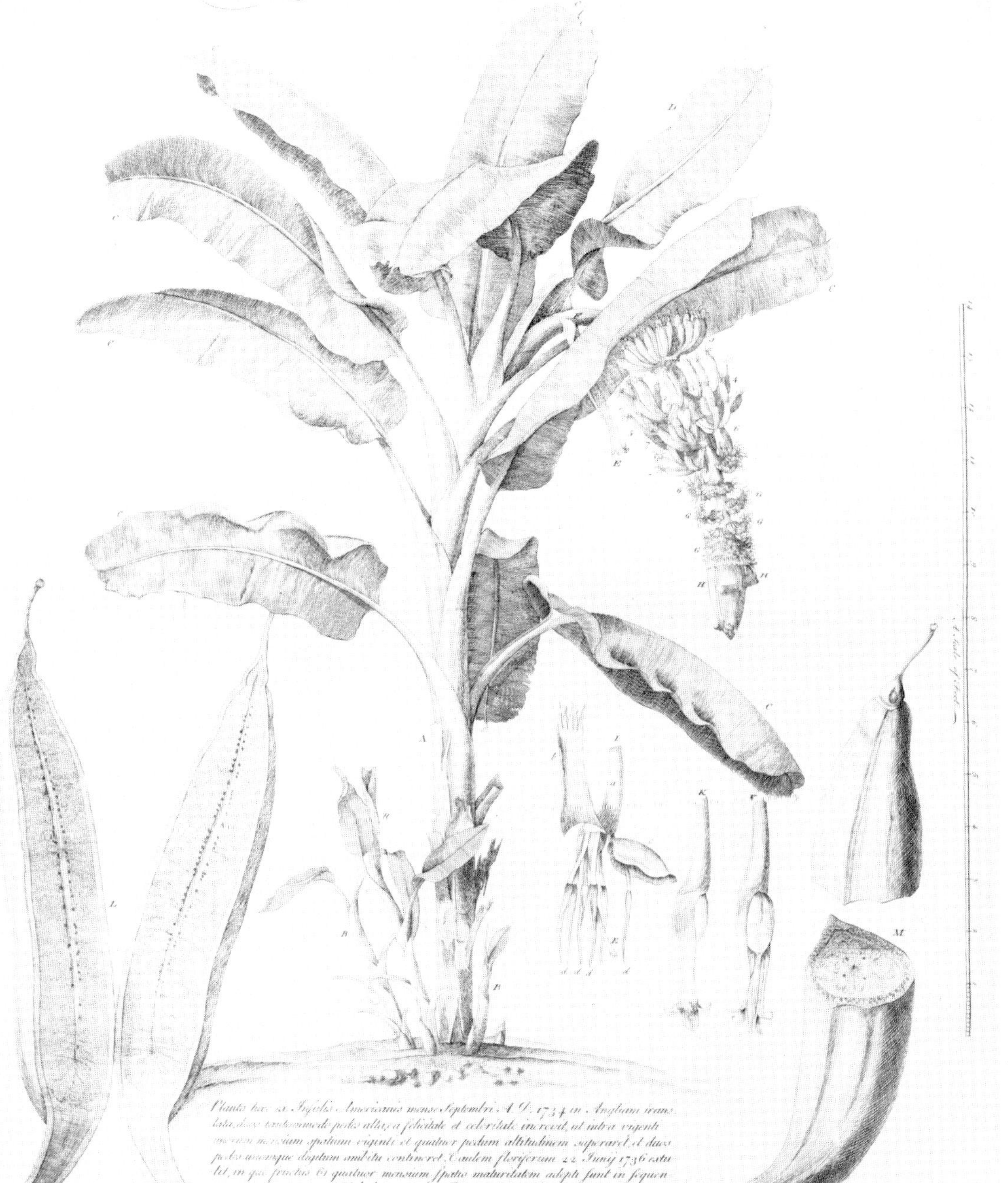

Planta haec ex Insulis Americanis mense Septembri A.D. 1734 in Angliam translata, duos tantummodo pedes alta, ea felicitate et celeritate increvit, ut intra viginti novem mensium spatium viginti et quatuor pedum altitudinem superaret, et duos pedes unumque digitum ambitu contineret. Caulem floriferum 22 Junij 1736 extulit, in quo fructus 61 quatuor mensium spatio maturitatem adepti sunt in sequenti Tabula continetur Plantae descriptio.

A. Exhibet Truncum e pluribus Bulbis conflatum.
B. Stolones. C. Folia.
D. Folium ultimum.
E. Involucrum Caulis floriferi.
F. Caulem floriferum.
1. Verticillum primum fructuum Maturescentium.
2. Secundum.
3. Tertium.
4. Quartum.
5. Quintum.
6. Sextum.
7. Septimum.
G. Verticillos florum masculinorum.
H. Spathas verticillos florum includentes, [illegible] deciduas.
I. Florem foemininum.
a. Germen. b. Petalum.
c. Nectarium.
d. Filamenta Staminibus Masculinis similia. e. Pistillum.
K. Florem Masculinum. L. Fructus maturi Sectionem Longitudinalem.
M. Fructus maturi Sectionem latitudinalem.

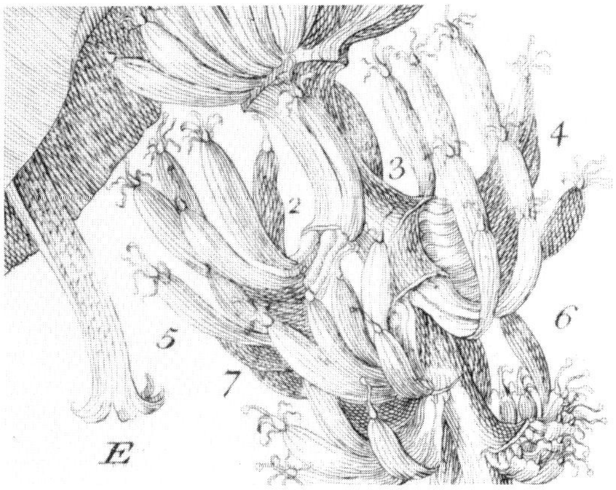

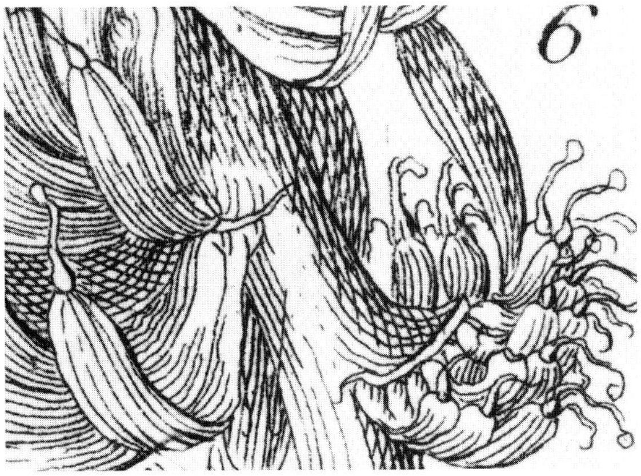

9. MUSA: FRUCTU CUCUMERINO LONGIORE. Line etching, cross-hatched shading, single bite, local burin finishing, with accompanying line-engraved text, etched by Georg Dionysius Ehret (German 1708-1770) after his own drawing. An illustrated broadsheet printed at London(?) ca.1736, plate printer unknown. Printed in black ink (platemark 63.5 x 46.5 cm.) on laid paper (wire side, sheet size 75.8 x 55.8 cm.). Separately published print probably issued by G. D. Ehret.

Etching had the particular convenience that it was swift and required little equipment. Someone skilled in the art could produce a reasonably simple plate and have it printed all in the space of a few hours.

G. D. Ehret, who occupied a dominant position in mid-18th-century botanical drawing, was also a competent etcher and prepared many plates from his own drawings. Early in 1736 he visited Holland and met the great Linnaeus, from whom he learnt the value of making floral dissections. Thereafter Ehret frequently provided a full floral dissection when illustrating any plant novelty, thus giving his already accurate plant portraits an added scientific value. In the summer of that same year he returned to England specially to record the exciting event of the flowering, in June, of a banana plant in Sir Joseph Ayloffe's garden. Apart from any painting he may have made, he used his skill in etching to prepare this large, annotated, visual record of the event, virtually a graphic botanical news sheet. It was produced as a separate publication, one of several similar prints issued by Ehret during his life.

Etching conveniently enabled him to sketch this record without undue delay so that copies could be sent to interested botanists such as his lifelong patron, C. J. Trew of Nuremberg. Ehret prepared the etching with minute attention to detail, adding elaborate line and cross-line shading, and included illustrations of the most carefully dissected detail. The plate was drawn mostly with a fine needled line and given a single "bite," but there is evidence of local strengthening of lines with the burin. The latter practice, known as "re-entering" the line, was yet another means of giving some calligraphic expression to needled lines. The formal calligraphic and other lettering, which is all burin work, was almost certainly done by a skilled "writing engraver" and it is possible that the same craftsman may have done some of the re-entering, which would have required special training in the use of the burin.

Very few copies of this, or any other of Ehret's separate prints, have apparently survived, and this one is not recorded in any source known to us. This particular copy apparently lay folded in some dusty situation for a great many years and probably came close to destruction after its novelty interest had declined.

Tab. XXVII.

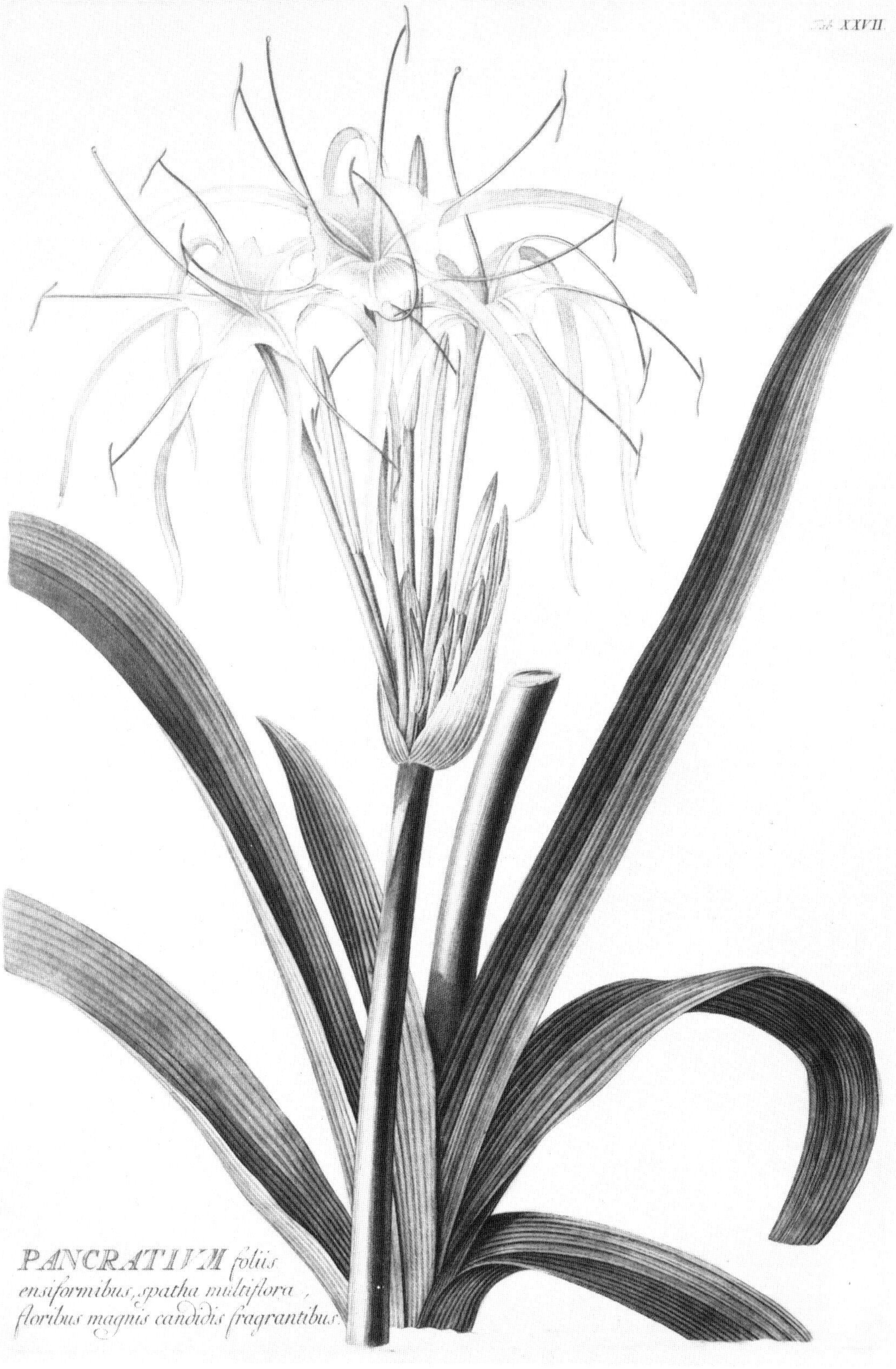

PANCRATIUM foliis ensiformibus, spatha multiflora, floribus magnis candidis fragrantibus.

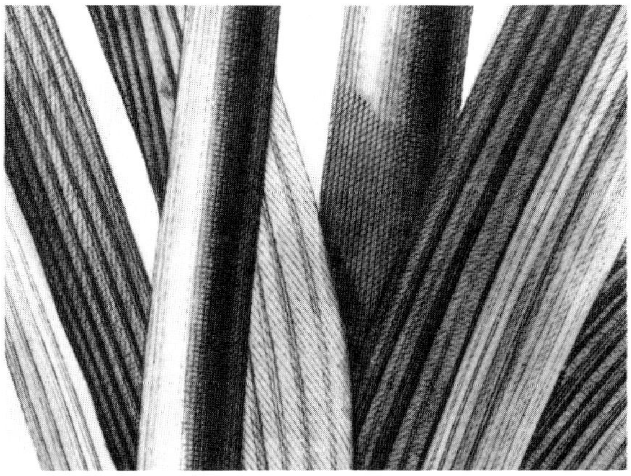

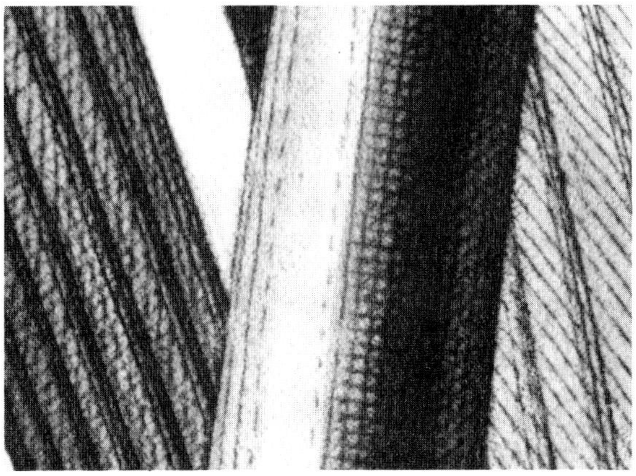

10. PANCRATIVM foliis / ensiformibus, spatha multiflora, / floribus magnis candidis fragrantibus. Line etching,with cross-ruled shading (2-3 bites), rocked-tint shading, probably engraved by Johann Jakob Haid (German 1703-1767) after a drawing by Georg Dionysius Ehret (German 1708-1770). Forming a full-page illustration printed at Nuremberg ca.1753 by J. J. Haid. Printed in black ink (platemark 45.6 x 30.2 cm.) on laid paper (top side, watermark of D. & C. Blauw, sheet size 51 x 32.6 cm.). Hand-finished with transparent watercolor. Ex: C. J. Trew. *Plantae Selectae. [Nuremberg, 1750-1773]. Plate 27. Nissen BBI #1997.*

Many of Ehret's flower portraits were engraved and published in the great botanical iconographies of his day. His patron, Trew, published 100 of them in his stately Plantae Selectae, *which commenced in 1750 and took 23 years to reach completion. Ehret's competent, but not particularly commendable, skill in etching was probably not equal to the standards demanded by the discerning bibliophiles of his day. The wealthy Trew engaged the Haids, a distinguished family of Nuremberg engravers, to translate Ehret's crisp, vigorously drawn paintings into printed illustrations using their own style of etching.*

Ehret frequently used body color for his painting and preferred to work on vellum. To give body to a line etching required some skillful technique in order to preserve the necessary modelling and perspective. The Haids employed a sophisticated style of what appears to be ruled line shading, with added cross-ruling for greater emphasis, and a second or third bite for stronger passages. Ruled shading has a somewhat mechanical effect, so delicate local shading was added with the mezzotint rocker. The whole was printed in black ink, but where paler effects were required—on the white flowers for instance—wiping was apparently very vigorous so as to give a very light impression. The finished plate was hand-colored with watercolor and the success of the engraver's modelling is made apparent, for the transparent color is given the effect of body where it runs over the shading.

All this skillful technique went to create one of the great botanical iconographies of the 18th century. Although engraved by other hands, its plates still carried the unmistakable stamp of Ehret's masterly style.

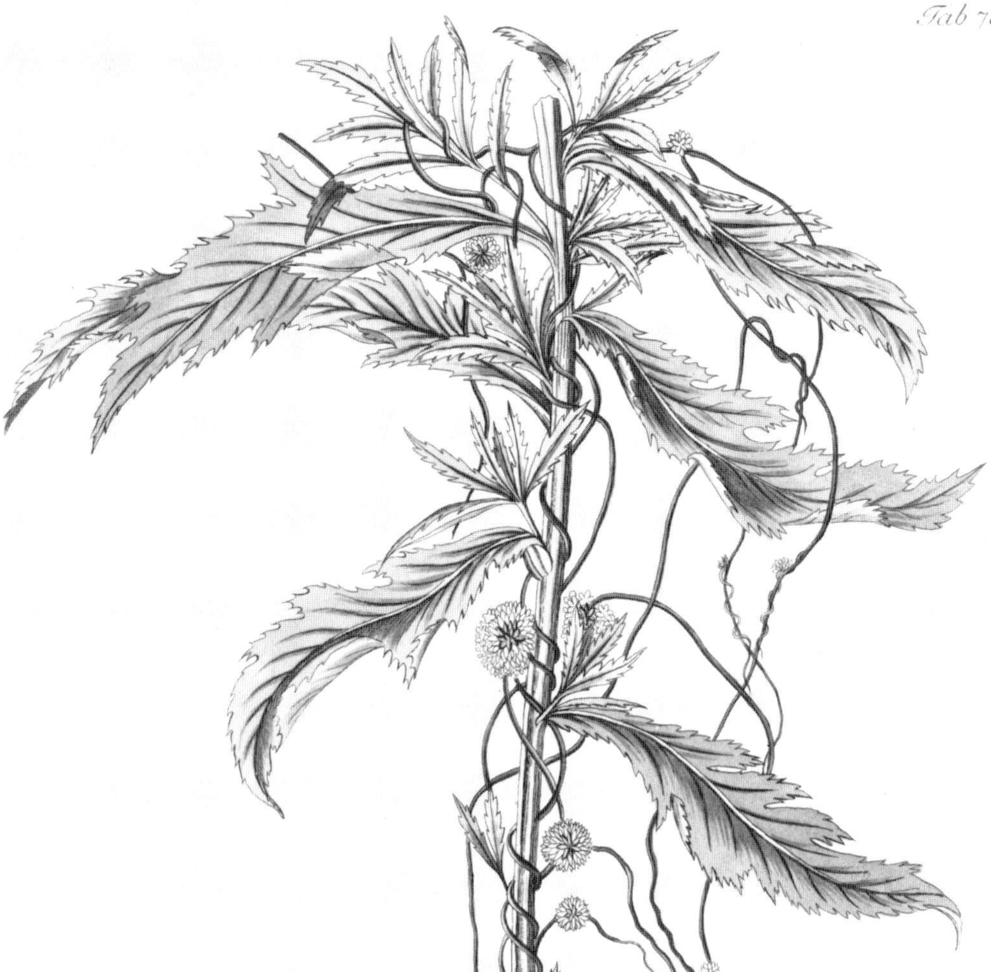

CUSCUTA EUROPÆA L.
Der europæische Leindotter.

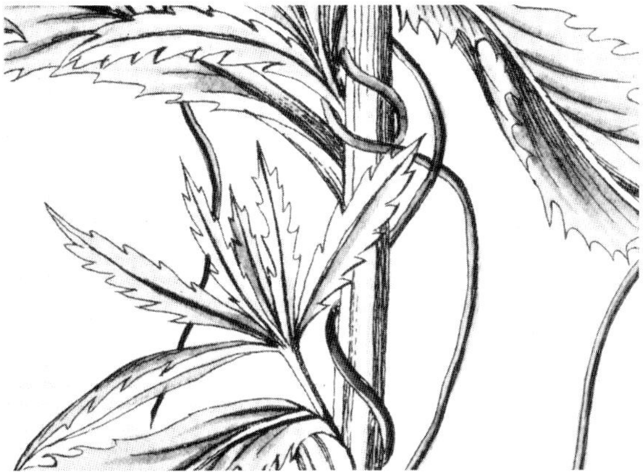

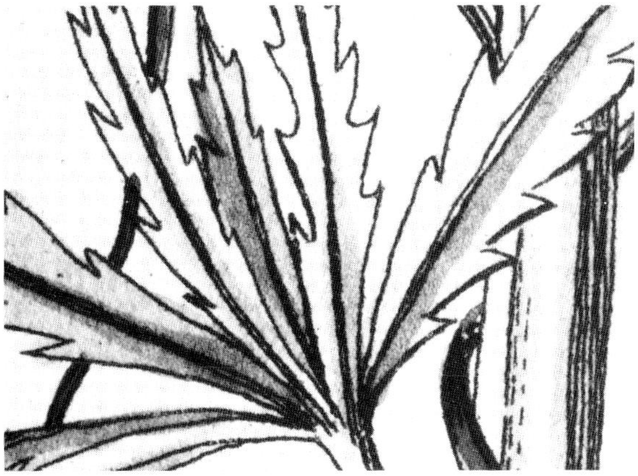

11. CUSCUTA EUROPAEA. L. / Der europaeische Leindotter. Line etching, single bite, etcher and artist unknown. Forming a full-page illustration printed at Vienna in 1788, plate printer unknown. Printed in black ink (platemark 38.3 x 25.8 cm.) on laid paper (top side, bluish-white color, sheet size 47.5 x 32.1 cm.). Hand-finished with transparent watercolor. Ex: J. J. von Plenk. *Icones Plantarum Medicinalium...* Vienna, R. Graeffer & Soc., 1788-1812. Plate 70. Nissen BBI #1536.

The elaborately prepared plates that the Haids etched after Ehret's stately flower paintings had a formality that placed them with the aristocracy of flower prints. But they had none of that spontaneity that is the special quality and virtue of etching. The Haids could afford to spend as many years as they did in preparing their plates, for the wealthy C. J. Trew was paying the bills.

Typical of much of the more commonplace use of the medium is the extensive iconography of medicinal plants published by Plenk in Vienna. No fewer than 758 plates were prepared for this work between 1788 and 1812. The unnamed artists and etchers maintained a high standard with vigorous, decorative and accurate illustrations of plants, all achieved with minimal sophistication, sound draughtsmanship and competent single-bite etching.

Etching was to prove the most accessible and inexpensive medium for the very large number of botanical prints published in the late 18th and early 19th centuries. The hundreds of illustrations in Curtis' Botanical Magazine *and the many comparable iconographies were almost all etchings. It was not to be superseded until the adoption of chalk-style lithography by such artist/lithographers as W. H. Fitch in the 1830s.*

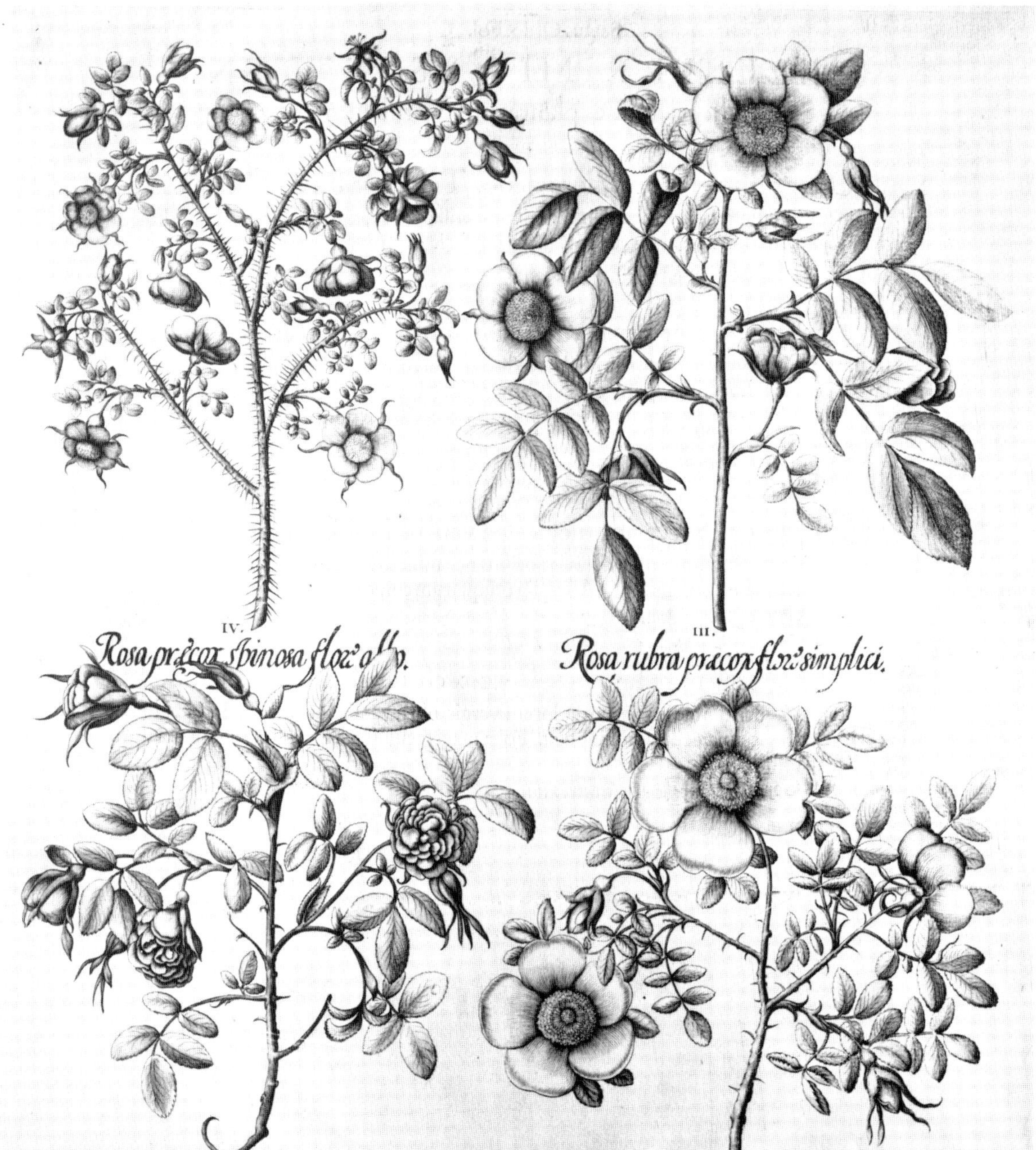
IV.
Rosa præcox spinosa flor: alb.
III.
Rosa rubra præcox flo: simplici.
II.
Rosa Cinnamomea.
I
Rosa lutea flore simplici.

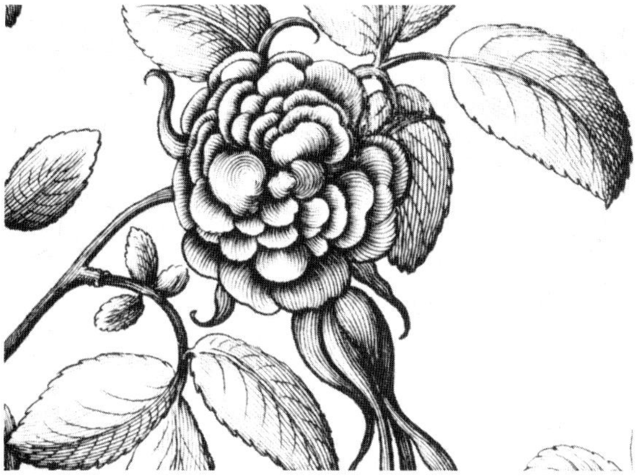

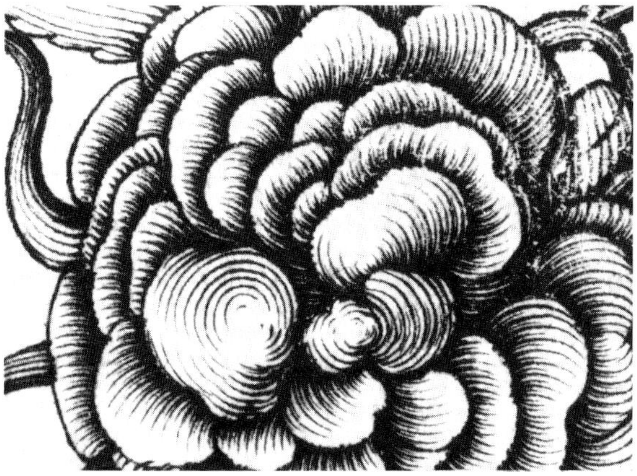

12. IV. Rosa praecox spinosa flore alb. III. Rosa rubra praecox flore simplici. / II. Rosa Cinnamomea. I. Rosa lutea a flore simplici. Line engraving, cross-hatched shading, pale ink tone, possibly engraved by Wolfgang Kilian (German 1581-1662) after drawings by Basil Besler (German 1561-1629). Forming a full-page illustration printed at Nuremberg in 1613, plate printer unknown. Printed in black ink with slight ink tone (platemark 47.0 x 39.4 cm.) on laid paper (wire side, sheet size 52.7 x 43.1 cm.). Ex: B. Besler. *Hortus Eystettensis.* (Sextos ordo: Collectarum plantarum vernalium.) [Eichstätt and Nüremberg], 1613. Vol. 1, folio E6r. Nissen BBI #158.

Parallel with the development of etching in botanical printmaking was that of line engraving. Line engraving was the antithesis of the generally free and unlaboured process of etching. It was essentially a formal, laboriously achieved and highly sophisticated medium with a much fuller visual syntax than that of etching. However fine and apparently clean an etched line seems, it has a somewhat ragged edge when seen under magnification because the mordant eats the copper in an irregular way. In comparison the furrow sculpted in the plate by the keenly sharpened, precisely vee-shaped burin can have a knife-edged sharpness and a cold exactness that has no equivalent in etching. Furthermore, the production of lines with the burin requires a firmly controlled muscular effort which gives them a purposiveness and force quite different from the freedom of lines drawn with the needle, which merely skates the surface.

Besler's mighty iconography illustrated over 1000 plants on 374 spacious plates. It is the work of ten line engravers of whom the chief, W. Kilian, had been trained in the Italian broad manner, that is, using swelled lines that share the character of calligraphic penmanship. The somewhat flat and stylized depiction of Besler's plants is achieved with coarsely engraved lines supplemented by close and carefully laid hatched and cross-hatched shading. Some plates are arranged in quartets, like the example shown, but many others are life-sized engravings that fill the page.

Thomas Johnson, editor of the 1633 edition of Gerard's Herball, *referred to Besler and wrote of "the figures being very large, and all curiously cut in brasse..." Brass was an unusual metal to use for engraving, copper being almost universally preferred. However, the plates were available for reprinting as late as 1713 and Johnson may have seen the plates for himself. This record of plants growing in an Eichstätt garden, arranged by season and presented in imposingly dramatic and decorative compositions, is a justly famous landmark in early botanical printmaking.*

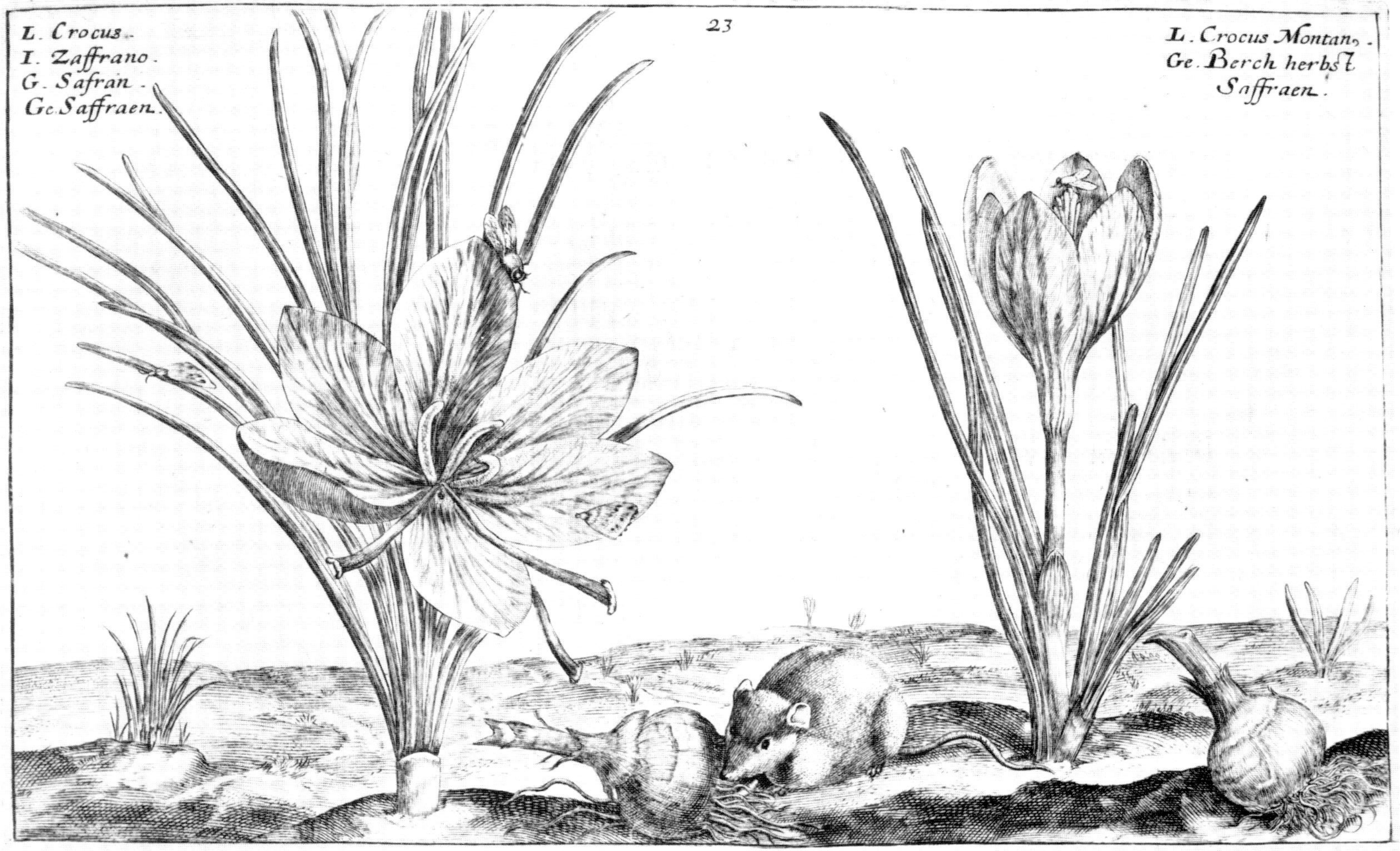
23
L. Crocus.
I. Zaffrano.
G. Safran.
Ge. Saffraen.
L. Crocus Montan.
Ge. Berch herbst
Saffraen.

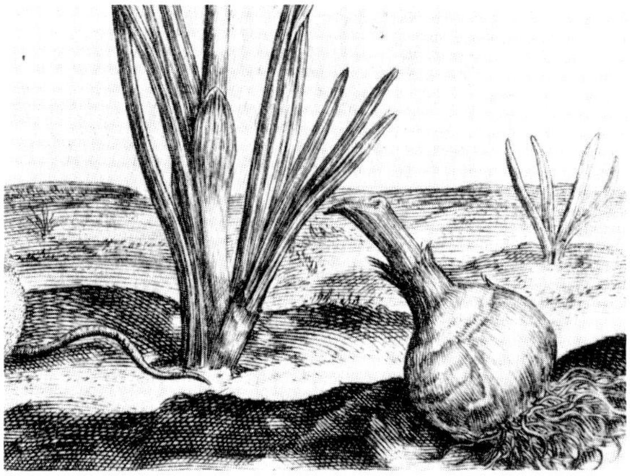

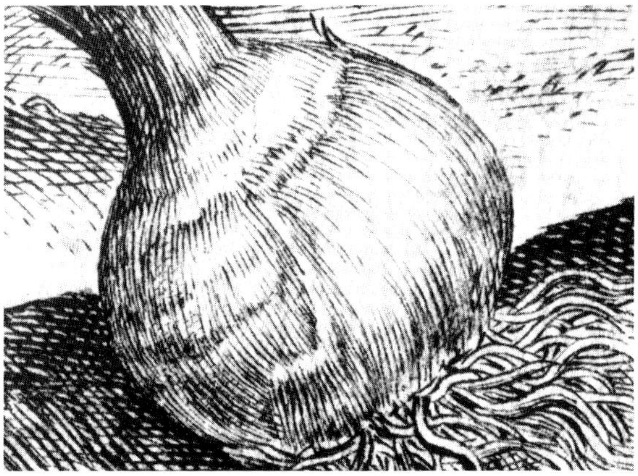

13. CROCVS SATIVVS. ET MONTANVS PRIMVS. (title on facing text page). Line engraving, cross-hatched shading, engraved by Crispijn van de Passe (Dutch 1593/94-1667) probably after his own drawing. Forming a full-page illustration printed at Utrecht in 1616 by Officina Caelatoria C. Passe. Printed in black ink (platemark 13.7 x 21.2 cm.) on laid paper (top side, page size 17.4 x 25.3 cm.). In: C. van de Passe. *Hortus Floridus*. Utrecht, C. de Pas; Arnhem, J. Janssoon, [1607?-]1614[-1617]. Part 3: Autumnus, plate 23. Nissen BBI #1494.

At about the same time that Besler's great folio was going through the press, a distinguished Dutch line engraver was preparing a very different seasonal flower iconography. With smallish, mostly landscaped plates of great delicacy and polish, typical of Dutch work of the day, Crispijn van de Passe's florilegium has won the affection and admiration of botanists and artists ever since.

The engraver used some of the classical language of line engraving to give his works that sort of polish. The commencement and completion of lines engraved with the burin usually have an acute sharpness, and a line may swell and diminish in width along its length. This latter was achieved by ploughing deeper into the copper with the vee-shaped burin, which naturally widened the furrow as it deepened it. Besler's engravers made particular play of this language but Passe engraved a much more restrained line, preferring to mass his lines into closely laid shadings with free use of cross hatching. When more delicate passages demanded it, he replaced actual lines with mere flicks of the burin.

The overall effect is of refined elegance with pronounced attention to modelling and perspective. Like so many Dutch artists, Passe favoured a very low perspective such as we see here, where we find ourselves sharing the mouse's view of autumnal crocusses. Typically, the plants are shown growing, rather than uprooted as the earlier herbal artists preferred, but in order to include information about the roots some bulbs have been strewn on the surface of the soil. Also typical of this work is the presence of moths and flies and the homely field mouse—all calculated to win the attention and favor that the artist has always received.

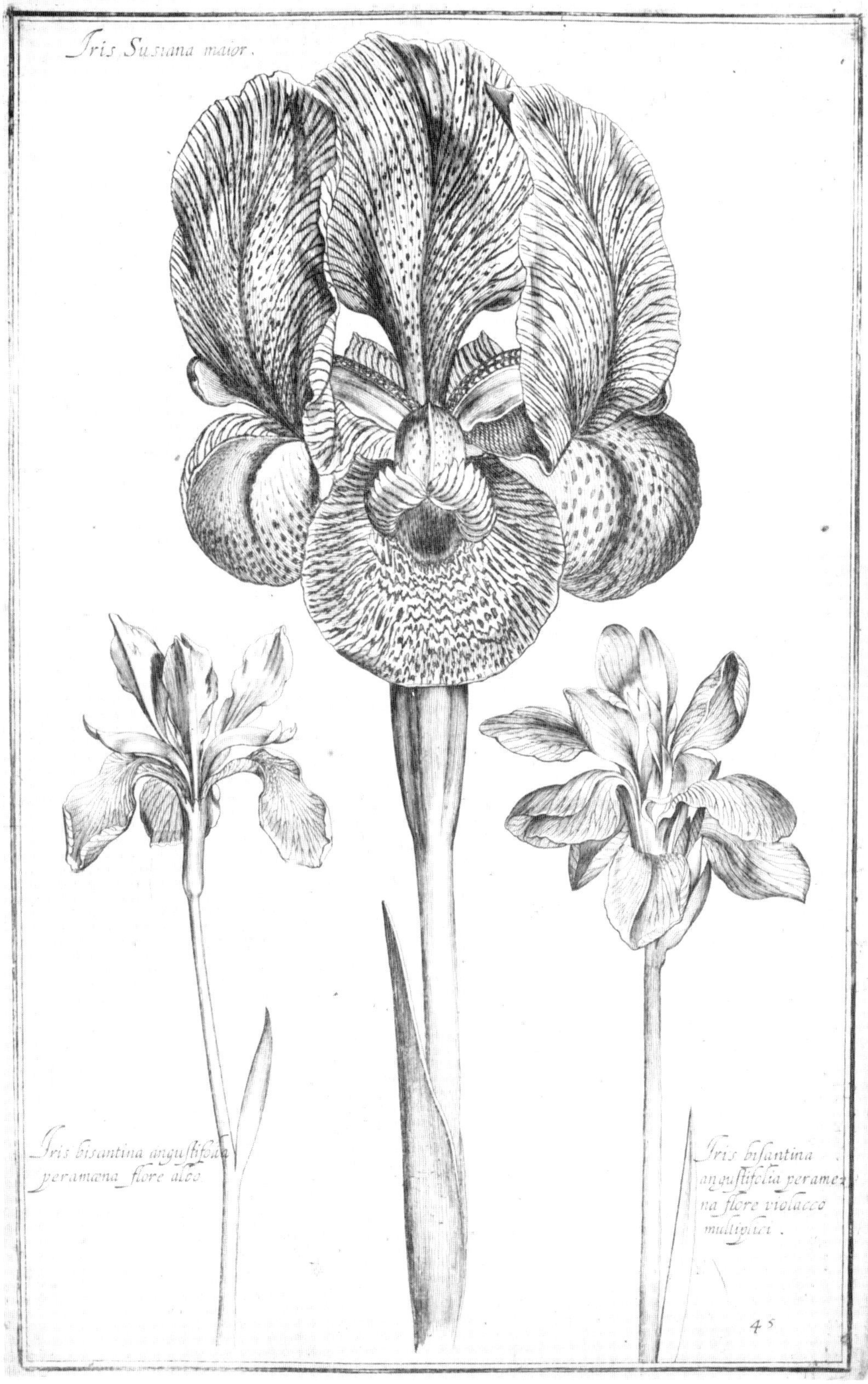
Iris Susiana maior.
Iris bisantina angustifolia
peramæna flore albo
Iris bisantina
angustifolia perame-
na flore violacco
multiplici.
45

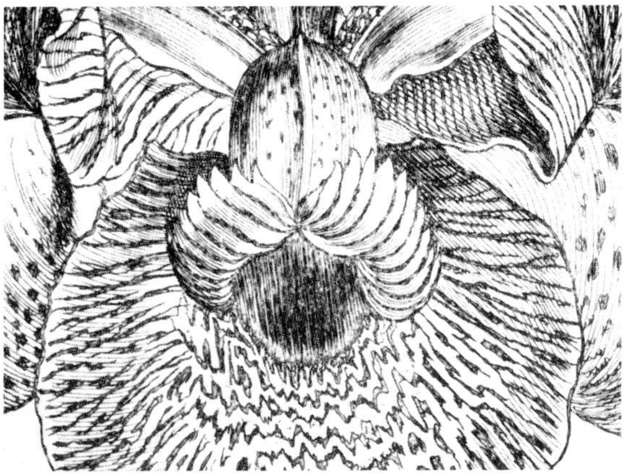

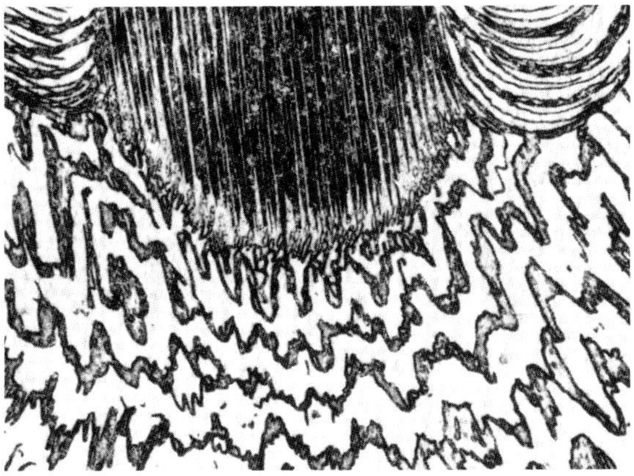

14. Iris Susiana maior. Iris bisantina angustifolia / peramaena flore albo. *and* Iris bisantina / angustifolia perame: / na flore vialacco multiplica. Line engraving, with much local etching, engraved by Daniel Rabel (French 1578-1637) after his own drawing. Forming a full-page illustration printed at Paris in 1622 by N. de Mathonière. Printed in black ink (platemark 33.8 x 22.2 cm.) on laid paper (wire side, sheet size 35.4 x 23.3 cm.). Ex: [D. Rabel]. *Theatrum Florae.* Paris, N. de Mathonière, 1622. Plate 45. Nissen BBI # 1575.

The presentation of portraits of flowers that placed greater emphasis on their beauty or novelty than on mere botanic interest gave rise to the class of books known as "florilegia." In these works text is often reduced to a subsidiary rôle, and is sometimes absent, there being no more than the legend engraved on each plate.

Rabel's Theatrum Florae *is just such a work. It presents a gallery of 69 engraved translations of his superb transparent watercolors, which Blunt considered "one of the marvels of flower painting." Rabel was apparently his own printmaker and the results, though highly satisfying in their own right, apparently do not win everyone's favor, for Blunt wrote of "the immense superiority of Rabel's paintings over his engravings." That comparison can still be made, for the original watercolors survive in the Bibliothèque Nationale in Paris. Savage, looking at the work from another point of view, suggested that "The increasing use of copper-plate engravings as botanic illustrations at the end of the sixteenth and the beginning of the seventeenth centuries proved a new incentive to good botanic painting, as the engraver could then do fuller justice in reproducting the light and shade of delicate painted work."*

As the son of a painter and engraver, and proficient in both arts himself, Rabel was obviously quite aware of the possibilities of both media. He was a consummate colorist but also had a keen sense of the tonalities of the print. He performed an extraordinary feat in translating the colors of iris petals in the black and white portraits shown here. This is achieved by the use of ordinary, but careful, engraved line and a highly individual etching technique for the "color" work. Apart from their rich suggestion of color, his engravings convey a convincing plasticity. Their upright, majestic formality demands our respect rather than the affection that Passe's intimate homeliness strove to win.

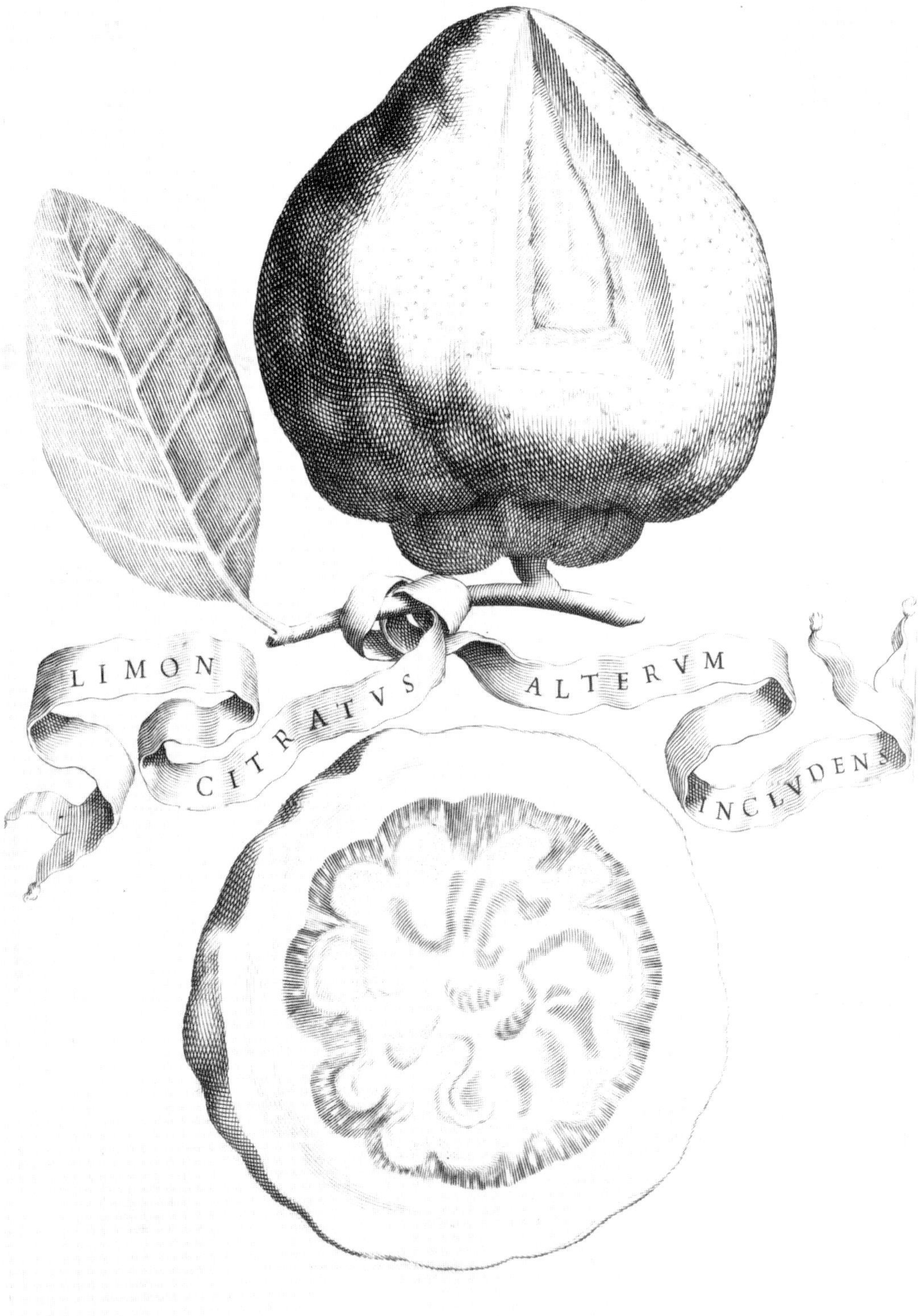
LIMON CITRATVS ALTERVM INCLVDENS

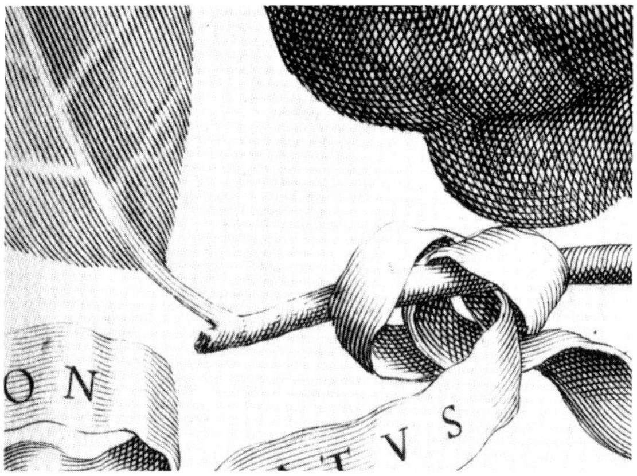

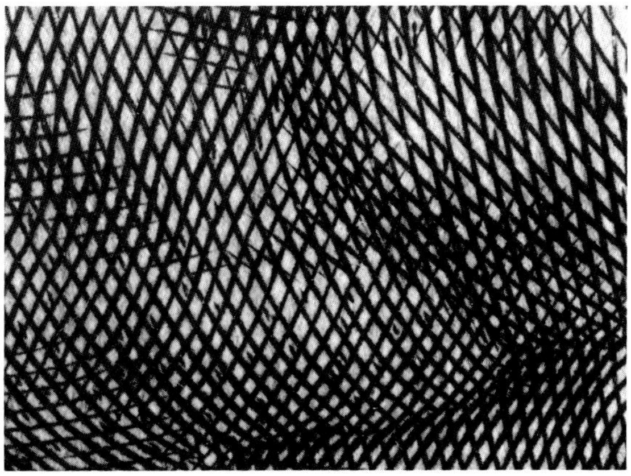

15. LIMON CITRATVS ALTERVM INCLVDENS. Line engraving, with cross-hatched and double cross-hatched shading, probably engraved by Cornelis Bloemaert (Dutch 1603?-1683) after his own drawing. Forming a full-page illustration printed at Rome in 1646, plate printer unknown. Printed in black ink (platemark 30 x 19.9 cm.) on laid paper (wire side, page size 35 x 23.5 cm.). In: G. B. Ferrari. *Hesperides*. Rome, H. Scheus, 1646. Page 269. Nissen BBI #621.

The portrayal of flowers and foliage is obviously the major concern of the botanical printmaker but there is, of course, a similar need to illustrate fruits in all their species and varieties. Some iconographies deal with fruit alone, of which a curious early example is Ferrari's Hesperides. *Apart from the portraits of fruit,* Hesperides *is embellished with several handsome allegorical plates engraved in a most accomplished style. Most of these are by the hand of C. Bloemaert, one of a school of craftsmen with an eloquent command of the formal language of the burin.*

With this single exception the rest of the 79 plates of fruit are etchings done by artists of varying degrees of competence under Bloemaert's supervision. This, the only line-engraved fruit plate, seems to bear the stamp of Bloemaert's masterly style and is almost a model exercise in certain aspects of burin technique.

The whole fruit, the sliced section and the leaf are all engraved without any outline. The leaf is entirely created with parallel swelled lines whose modulations convey form, venation, and to some extent, texture. The whole fruit provides a splendid example of the exploitation of hatched, cross-hatched and even double cross-hatched shading. Its texture is suggested by dextrous groups of burin flicks while different kinds of line shading depict the varied substances within—the latter revealed through an aperture that has apparently been chiselled through the side! The sectioned fruit again demonstrates how dots, flicks, swelled lines and parallel shading can be used, though the result has more of the coldness of polished agate than the juicy substance of a fruit. The accomplished lettering on the studied flutterings of the silken streamer complete this austere exercise in engraving.

One can almost sense the bright glitter of each meticulous furrow as the engraver ploughed his burin forward with loving attention to every millimeter of its path. Such a tour de force *of line engraving forms a total contrast to the sprightliness of etching as exemplified by Colonna's little prints (item #7) made only 30 years earlier.*

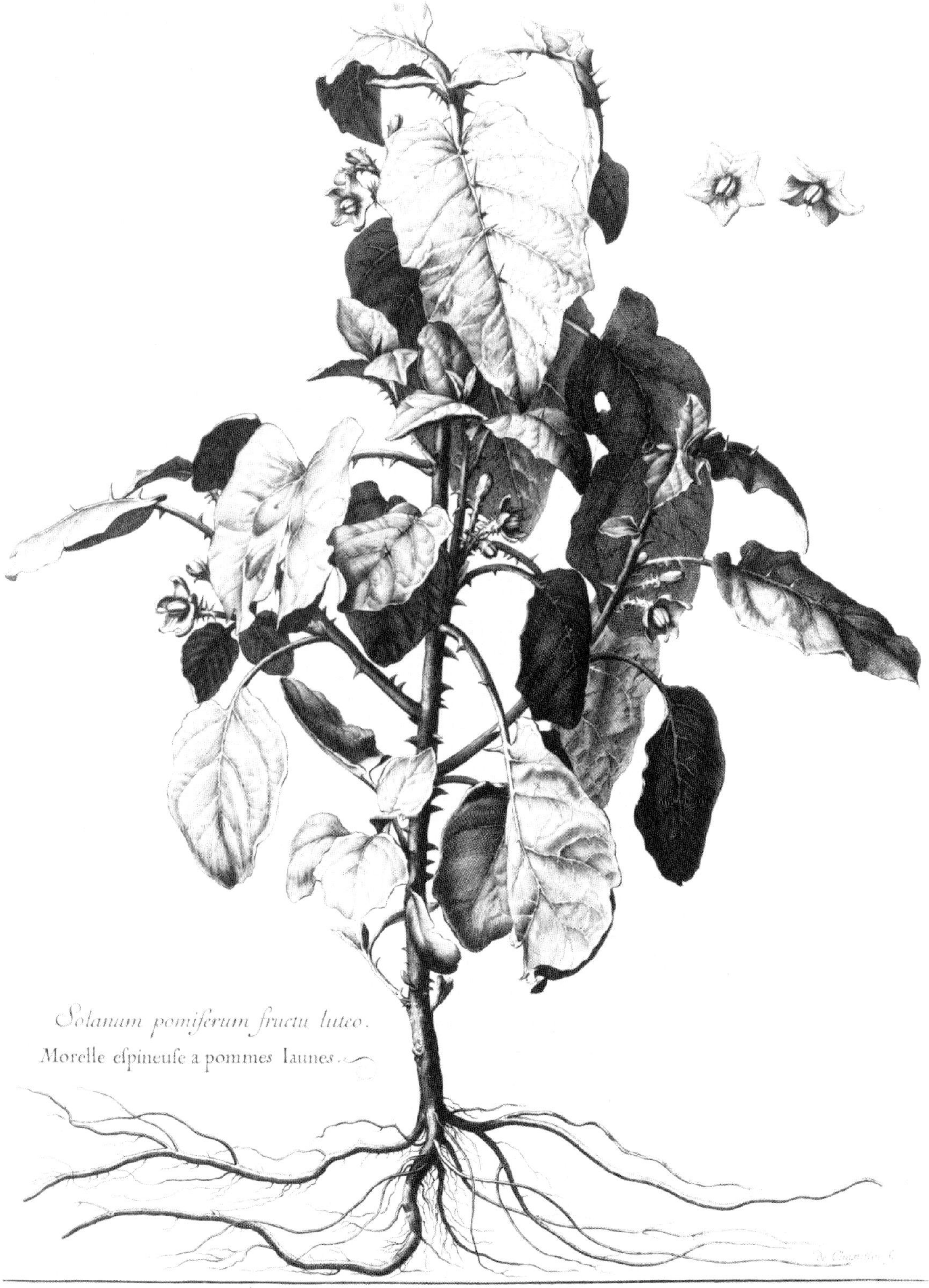

Solanum pomiferum fructu luteo.
Morelle espineuse a pommes Iaunes.

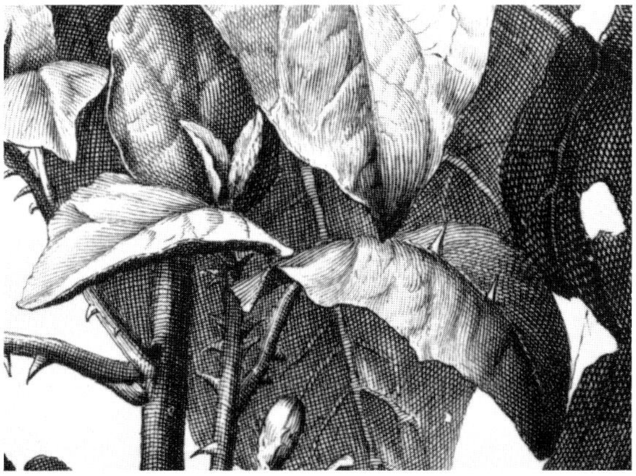

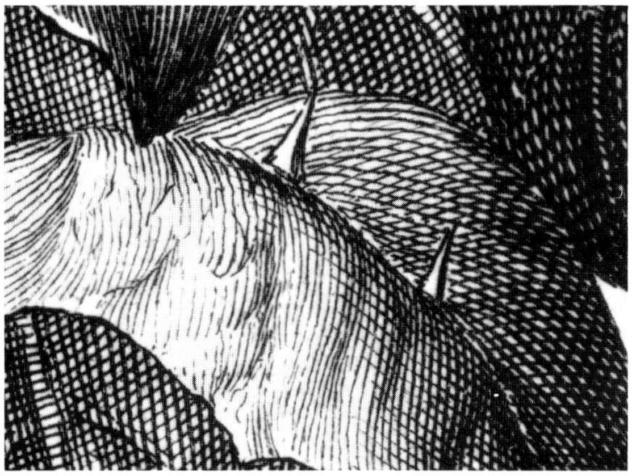

16. Solanum pomiferum fructu luteo. / Morelle espineuse a pommes launes. Line engraving, with some etching, engraved in the late 17th century by Louis Claude de Chastillon (French 1639-1734) after a drawing by Nicholas Robert (French 1614-1685). Forming a full-page illustration, this restrike printed from the 17th-century plate at Paris in recent years by Musée Louvre Chalcographie. Printed in black ink (platemark 42.6 x 31.3 cm.) on wove paper, (top side, sheet size 56.5 x 38 cm., with blind stamp of Musée Louvre Chalcographie). Prepared for the unpublished continuation of: D. Dodart. *Mémoires pour Servir à l'Histoire des Plantes.* Paris, Impr. Royale, [1675-]1676. Nissen BBI #502, viz [*Estampes pour Servir à l'Histoire des Plantes.* Paris, 1701. 319 pls.].

Without doubt some of the most elaborately conceived botanical line engravings are those prepared for the great botanical iconography that the French Académie Royale des Sciences proposed to publish under the editorship of Dionys Dodart. The first Mémoires *appeared in 1675, illustrated by 39 masterly plates, but the intended continuations never materialized. However, some 319 plates completed in anticipation were eventually issued without any text in 1701 and reprinted again in 1788. They were the work of three masterly printmakers, of whom L. C. de Chastillon was outstanding, as this example clearly shows.*

Dodart's preface to the Mémoires *clearly indicates that botanical accuracy was the foremost requirement. He directed that plants were to be portrayed in life size, shown in two such sections if not more than twice the page height; if even larger, then some portion of the plant was to be shown in natural size. Because of the shortcomings of colorists, the plates were intentionally presented in monochrome. However, Dodart directed that the engraving be skillfully modulated to indicate distinctions in the depth of color in different parts of the plant. Growth habit and structure, down to the minutest detail of texture and pubescence, were to be faithfully portrayed, and many plates included an additional illustration of the seedling stage.*

All this information was conveyed in the syntax of line engraving with discreet local use of etching. In fact, pure line engraving as used in Bloemaert's lemon is quite uncommon. From the early 16th century onwards, engravers used etching either as a means of strengthening burin lines or else, mixed with line engraving, to broaden the syntax of their graphic language.

Numerous authorities agree that Dodart's plates rank among the best botanical engravings ever produced. Chastillon's prints, in particular, are singled out by Blunt for their "exquisite delicacy." His original drawings for these plates are preserved at the Muséum National d'Histoire Naturelle in Paris. They are, Blunt tells us, "executed in sanguine on tissue paper...and are remarkable for their beauty and freshness." Most botanical engravers worked from colored drawings. It is revealing to discover that in this case the subtle planning of all color values, texture and shading was apparently worked out in monochrome before being translated into the language of engraving.

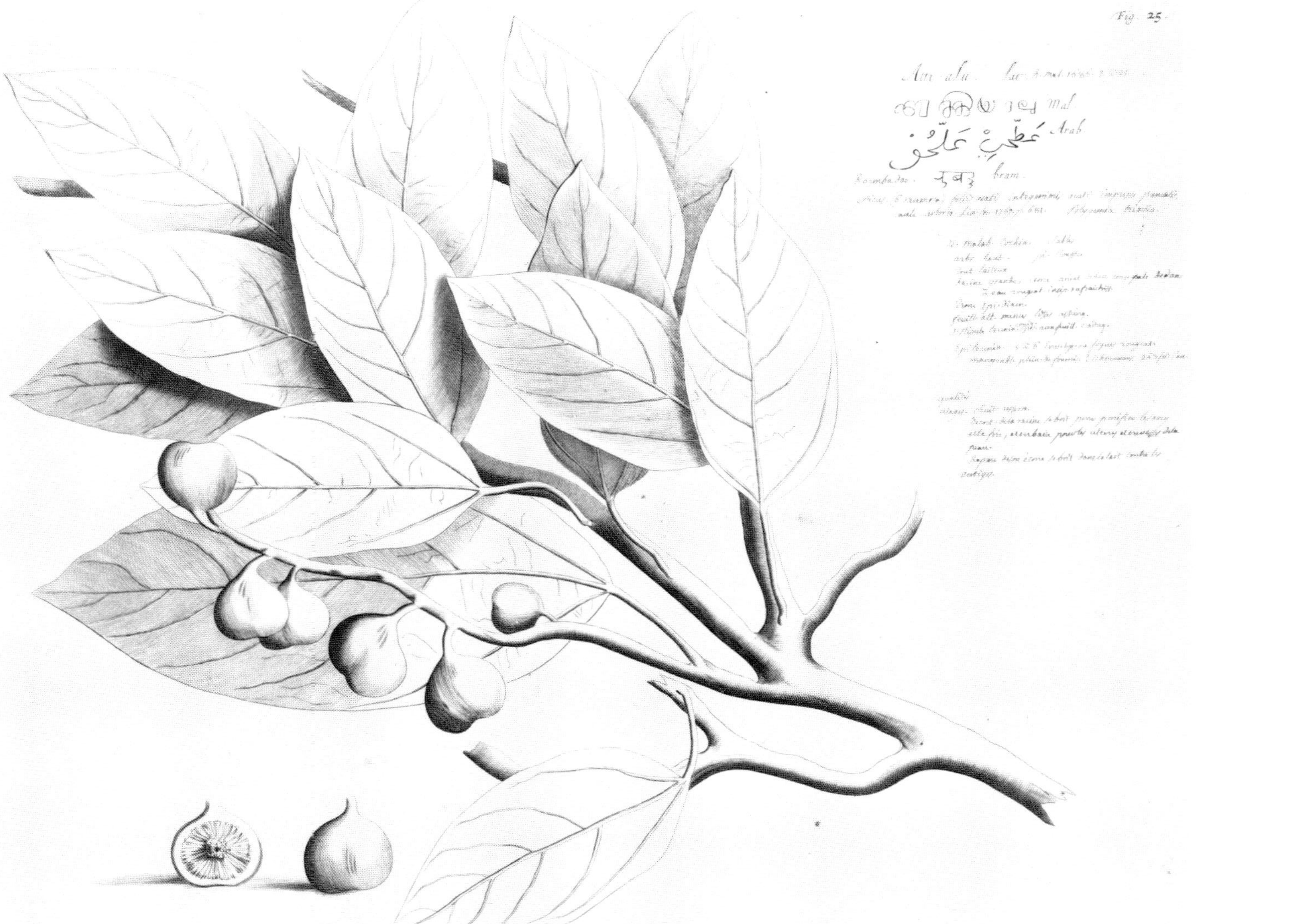

Fig. 25.
Mal.
Arab
bram.

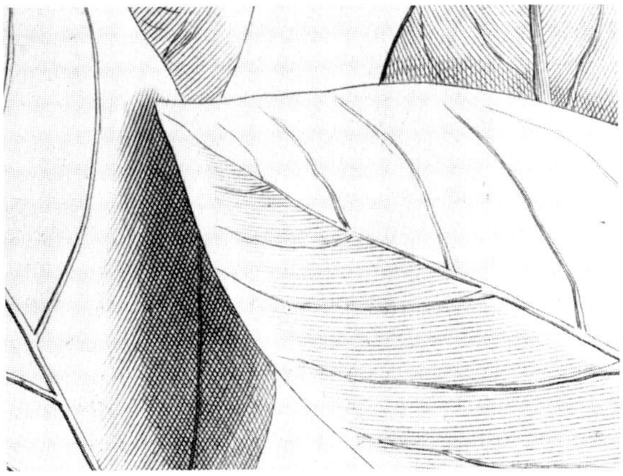

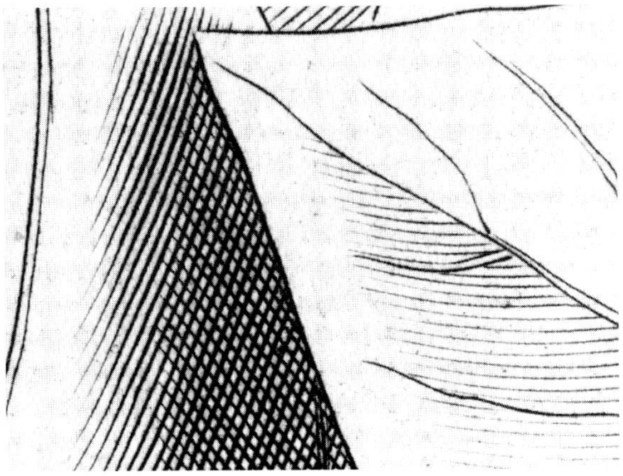

17. Atti-alu. Lat. ...Line engraving, outline style, with local cross-hatched shading, possibly engraved by Bastiaen Stoopendael (Dutch 1636/37-1707), possibly after a drawing by Pietro Foglia (i.e. Father Matthieu) (Italian ca.1617-1691). Forming a double-page illustration printed at Amsterdam in 1678, plate printer unknown. Printed in brownish-black ink (platemark 34.2 x 44.5 cm.) on laid paper (top side, sheet size 38.3 x 49.2 cm.). Print annotated by Michel Adanson. Ex: H.A. van Rheede tot Draakestein. *Horti Malabarici*. Amsterdam, J. van Someren, 1678-1703. Volume 1, plate 25. Nissen BBI #1625.

Three years after Dodart published his superb engravings, a somewhat archaic but, botanically speaking, more important iconography was launched in Utrecht. Rheede's vast Hortus Malabaricus *was to run to over 790 plates and includes many line engravings done in the curiously open style seen here. Its manner of engraving is reminiscent of Bloemaert's work of 30 years earlier but the style of presentation presages the bold etched outlines of Plumier's plant portraits of the 1690s (item #8).*

The plant is presented in an unusually flattened perspective with every leaf turned full-face to the viewer as though it was a pressed specimen. The gross structure of the leaves is delineated in cold outline with starkly drawn indications of venation, but surface texture or modelling are scarcely suggested. Both branch and fruits are modelled with highly disciplined parallel and cross-lined shading, but again surface texture is not indicated. An attempt was made to relieve the flatness of the perspective by creating shadows with studiously academic line work, but, somewhat confusingly, this suggests more than one light source.

Nissen feels that the plates have an unmistakable oriental style, and, indeed, that may account for the curiously artificial presentation of the growth habit. However, it is the brittle hardness in the engraving that destroys any life in the plant.

Rheede tells us that several artists contributed the drawings, while other sources ascribe much of the credit to P. Foglia (Father Matthieu in Holy Orders), who both drew and engraved many plates. We can only guess at the artist who made these portraits and we have little help with discovering their engraver's identity either. The engraver B. Stoopendael signed the first plate in the book and his manner shares some similarity with the syntax of this print.

Despite their apparent shortcomings, these 794 plant portraits were immensely valuable to science, providing depictions of scores of oriental plants that were quite new to botanists of the day.

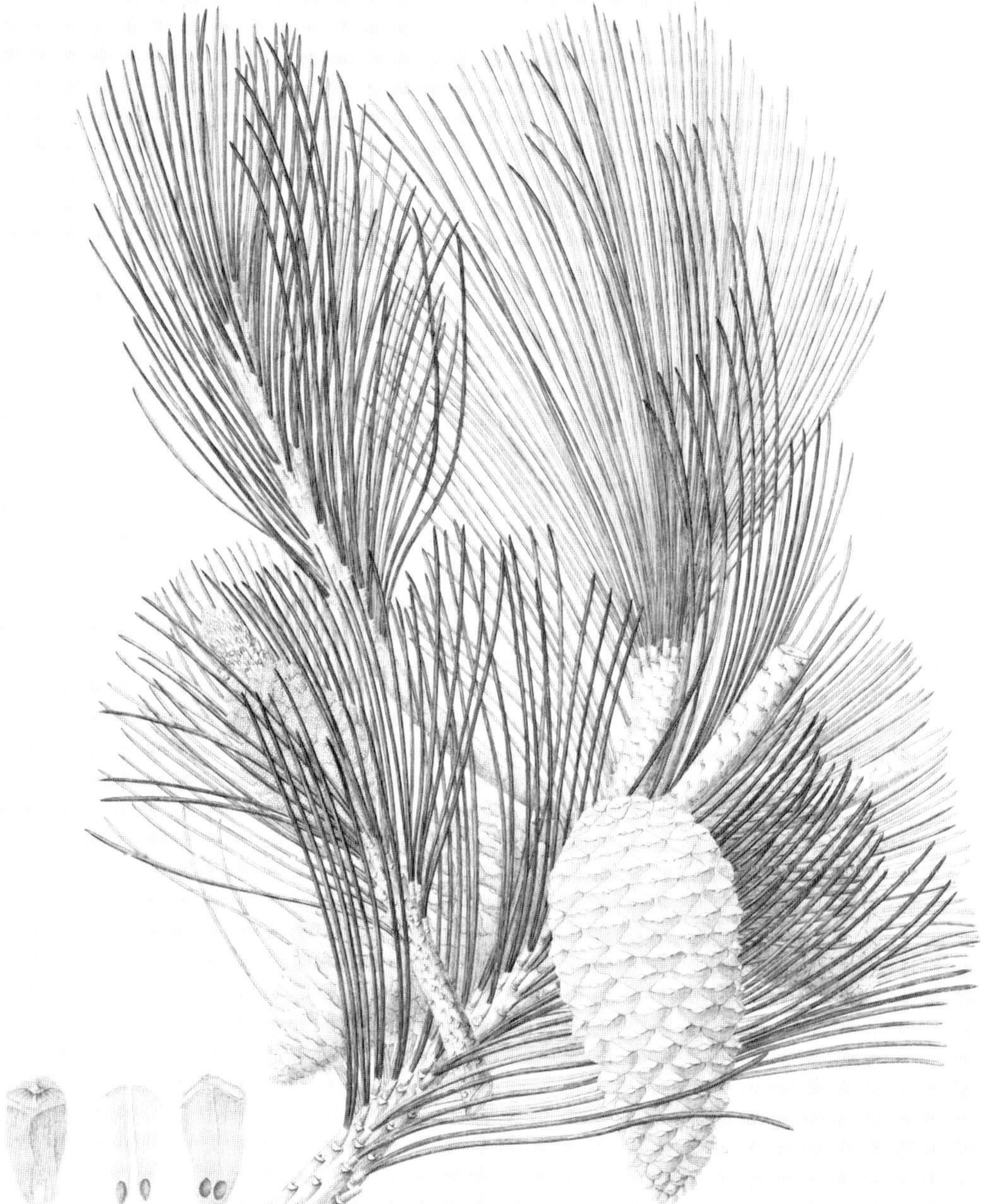

Pinus Pinaster

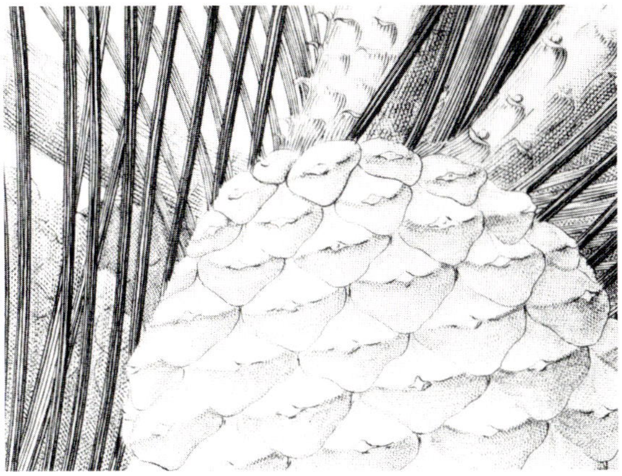

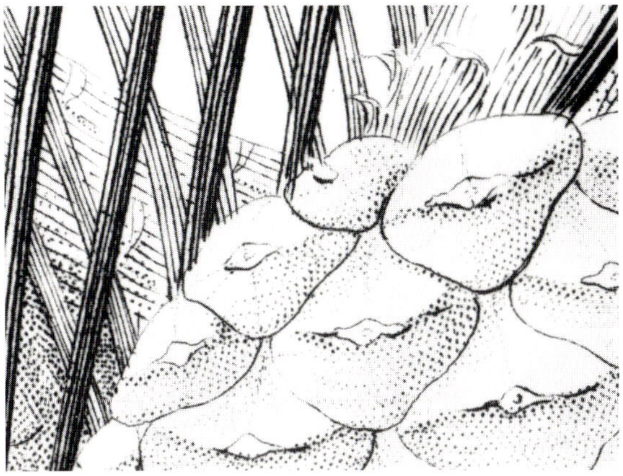

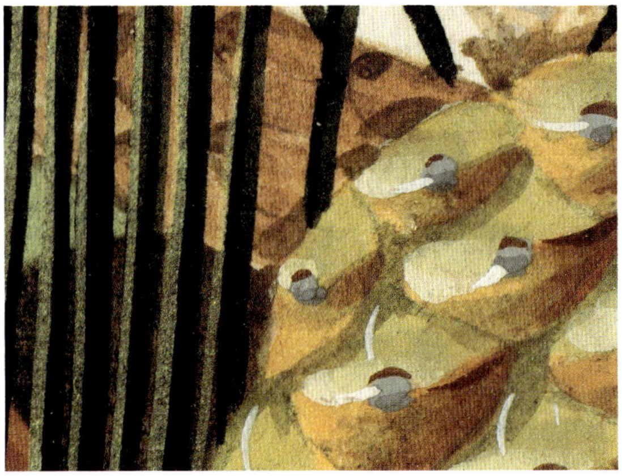

18. Pinus Pinaster. Line engraving, with stippled and roulette shading, engraved by Daniel Mackenzie (English fl. ca.1784-ca.1800) after a watercolor by Georg Dionysius Ehret (German 1708-1770). Printed at London in 1803, plate printer unknown. Printed in black ink (platemark 55.3 x 39.5 cm.) on wove paper (top side, sheet size 57.7 x 44.6 cm.). Ex: Lambert, A. B. *A Description of the Genus Pinus*. London, J. White, 1803-1824. Plate 5. Nissen BBI #1123. Shown together with a watercolor drawing of the same image, signed "G. D. Ehret pinx. 1744" (image area 47 x 33.5 cm.) on laid paper (wire side, sheet size 54 x 38 cm.).

Contrasting sharply with the spare, open delineation of the broad-leaved plant by Rheede's engraver is this elaborate rendering of a needle-leaved gymnosperm by Mackenzie. Almost regardless of expense, it seems, Lambert engaged the best English craftsmen to prepare the fine plates for his monograph on the pines.

Most of the plates were drawn specially for Lambert's work by Ferdinand Bauer and others, but just this one illustration was based on a watercolor by G. D. Ehret dating from 1744. It was Ehret's custom (probably in response to clients' demands) to make identical copies of some of his works, and two "originals" of this drawing are known to exist. One is shown here; the other—acknowledged as the engraver's model—is preserved with Sir Joseph Banks' collections at the British Museum (Natural History). The engraved version differs from the original in minor details and has some additional drawings of dissections.

Mackenzie's work merits a high distinction in the field of line engraving but he is almost unknown except to botanists. The considerable problem of delineating, in complicated perspective, the detailed form of scores of individual pine needles, the texture of the woody branches, and the form and texture of male and female cones would seem to present an insuperable challenge to any engraver. All this had to be translated from Ehret's watercolor drawing into the monochrome language of line engraving. Most is expressed in the familiar syntax of line engraving; each pine needle, for instance, is composed of five, six or seven closely laid lines running the entire length. A new element was introduced in the cones. They were modelled with dotted strokes made with the roulette—a toothed, freely rotating wheel mounted on a handle. Its special characteristics will be seen in a later exhibit (item #23). In comparison with Ehret's original this print actually contains more detailed information on niceties of minute form and texture. Clearly, a specimen of the plant must have been at hand to supplement the drawing and provide information on the dissected parts.

A very few copies of this book were specially hand-finished in watercolor by William Hooker. But, however skillfully applied, any coloring obscures the clarity of such closely laid engraving. Whilst adding new information to the print, it may be at the cost of losing the very detail that the engraver labored so carefully to include.

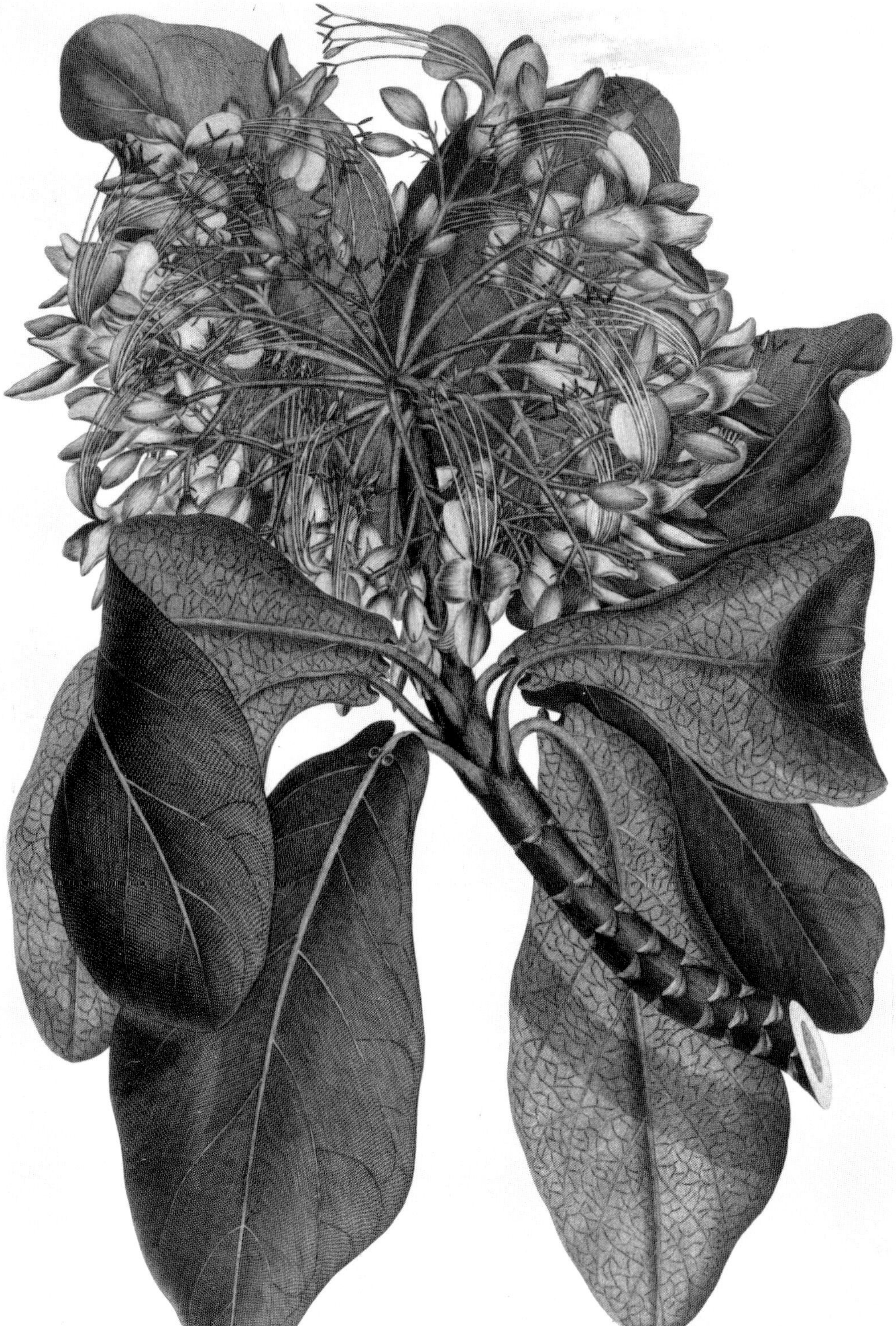

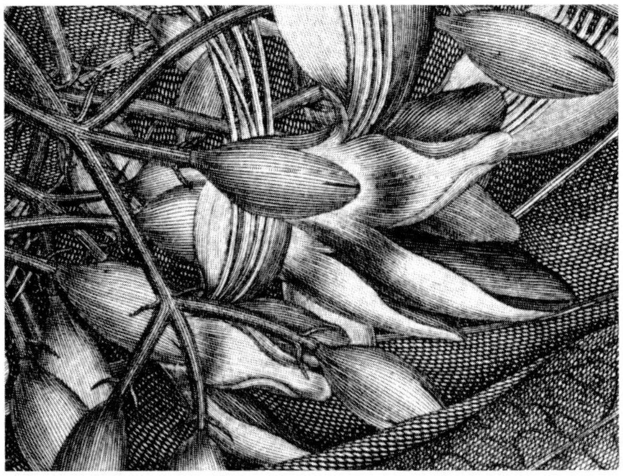

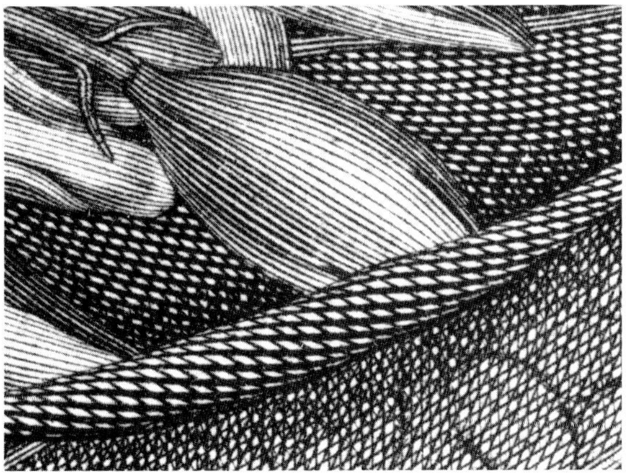

19. [Duplanthera tetraphylla Br.]. Line engraving, with densely laid cross-hatched shading, engraved in 1778 by Daniel Mackenzie (English fl. ca.1784-ca.1800) after a watercolor by Frederick Polydore Nodder (English?, fl. ca.1789-1800) based on a field drawing by Sydney Parkinson (English ca.1745-1771). Forming a full-page illustration, this restrike printed from the 18th-century plate at London ca.1966 by Thomas Ross. Printed in black ink (platemark 46 x 30 cm.) on Crisbrook Imperial 140-lb. H.P. wove paper (wire side, sheet size 57 x 39.5 cm.). Prepared for: [Sir J. Banks. Unpublished work on plants collected in 1768-1771 during Captain Cook's voyage round the world in H.M.S. "Endeavour"]. This restrike made for *Captain Cook's Florilegium*. London, Lion and Unicorn Press, 1973.

After his circumnavigation with Captain Cook on the "Endeavour" in 1768-1771, Sir Joseph Banks set about producing a great illustrated work to document the historic collections made on the voyage.

The surviving materials from Banks' enterprise include over 750 folio engravings, and more were planned; he had mentioned a likely total of 1100 in a letter to the younger Linneaus. Banks' ambitious and resourceful nature, supported by much personal wealth, led him to investigate the best printmaking techniques of his day: etching, stipple, line engraving, mezzotint, the newly discovered aquatint, and even à-la-poupée color printing. The powerful line engravings whose production he supervised emphatically proclaim his eventual choice.

They were prepared in the most highly elaborated style of line engraving, each one a virtuoso display of burin craft, with deep perspective and the minutest attention to every detail of structure. The coppers are still in existence at the British Museum (Natural History). The printmakers possessed comprehensive information by which to ensure the accuracy of their engravings; this comprised the actual plant specimens together with the field sketches, and highly finished studio paintings worked up from them. The vivid modelling and dramatic lighting of these prints are produced by densely laid nets of parallel and cross-hatched lines, all of which create a rich tonal spectrum of blacks and greys. Great attention was paid to the values *of tones so that, like Dodart's masterly plates of a century earlier, they would accurately convey monochrome distinctions of depth of color. This must have been calculated on the certainty that they would be printed in black, since the use of colored inks, inevitably lacking the force of black, would upset the balance of these values.*

Banks distributed specimen prints to some interested botanists of his day but, sadly, his great project died with him. However, these engravings, such grand demonstrations of the splendor of monochrome plant engraving, are a fitting memorial to his great vision, his choice of medium and his selection of craftsmen. This example, engraved by the industrious and skillful Mackenzie, was printed from the 18th-century copper some 20 years ago when a selection was published as Captain Cook's Florilegium.

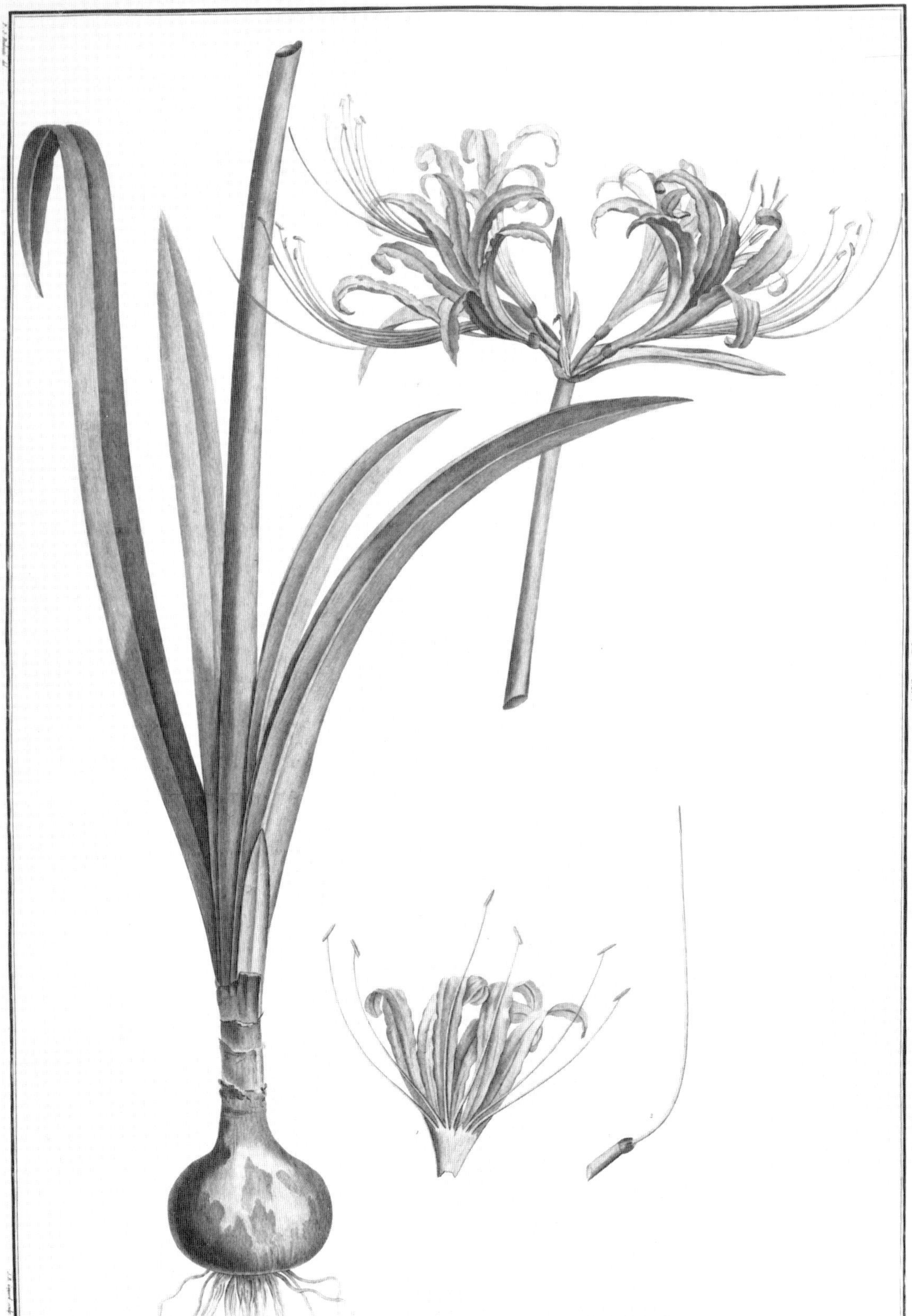

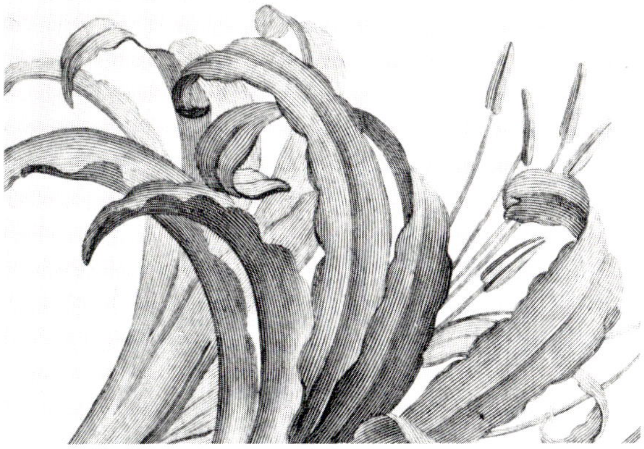

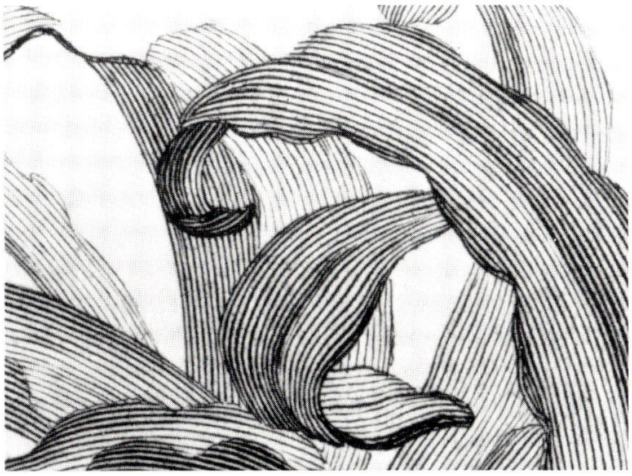

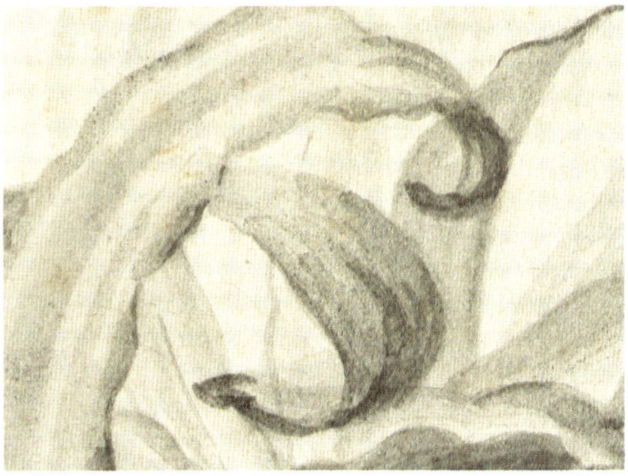

20. AMARYLLIS aurea. Line engraving, with parallel-line shading, engraved by Pierre-Joseph Redouté (Belgian 1759-1840) from a drawing by Jean Baptiste Guyard [Sr.] (French fl. ca.1788). Forming a full-page illustration printed at Paris in 1792, plate printer unknown. Printed in black ink (platemark 54.5 x 40.3 cm.) on laid paper (wire side, sheet size 62 x 48.2 cm.). Ex: C. L. L'Héritier de Brutelle. *Sertum Anglicum.* Paris, P. F. Didot, 1788[-1792]. Plate 15 bis. Nissen BBI #1189.

If Banks' plates were engraved without any intention of color printing, hand coloring was also quite out of the question. Transparent watercolor simply could not compete with the overall density of the image. Opaque colors, on the other hand, would have totally obliterated the engraver's work and almost required that the colorist paint a new picture over the print. These portraits were, indeed, supposed to be viewed as "black" prints. The highly finished full-color paintings previously prepared by Banks' artists were not intended as patterns for coloring. Rather, they served the valuable rôle of providing models for the engravers when constructing a scheme of tonal values with which to express distinctions in the depth of colors.

The 35 engravings made for L'Héritier's Sertum Anglicum *are of a humbler cast than Banks' sumptuous productions. They were competently line-engraved with scarcely any use of outline, employing overall parallel-line shading in a tonal manner to achieve modelling and perspective. Apparently L'Héritier always favored the special character of monochrome engraving for his various botanical publications. His artists prepared grey wash ("grisaille") drawings as patterns for these plates. Two of these originals are in the Hunt Institute collection, done by no less an artist than Redouté, so much identified with colored illustration in later years. Blunt (1963) remarks that they are "painted with the greatest care and attention to botanical detail, but they inevitably lack the subtlety and refinement of his more mature work and the brilliance that full colour always provides." However, they were eminently appropriate for this publication and the engravers successfully translated their subtle tones.*

Of the 35 engravings in this work, 31 are the first published illustrations of their subjects. As botanically educated and wealthy as L'Héritier was, it is remarkable that he should make no effort to provide color to prints that carried, and still carry, such importance for botanical science. Even at that date, in the middle of the classic period of botanical literature, it appears that color was not regarded as essential to the language of scientific illustration.

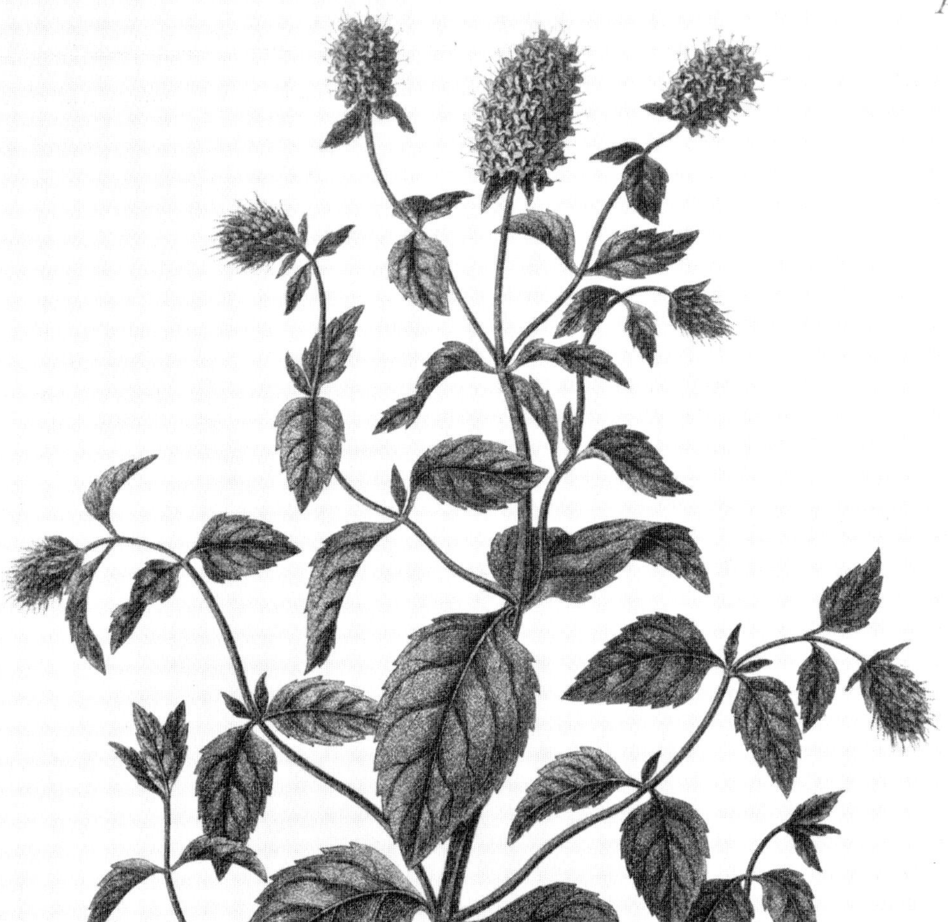

M: Piperita Vulgaris *Common Pepper-Mint.*

T. Sheppard, Del. *Bath, Pub: by W. Sole. 1797* Hibbert, S.

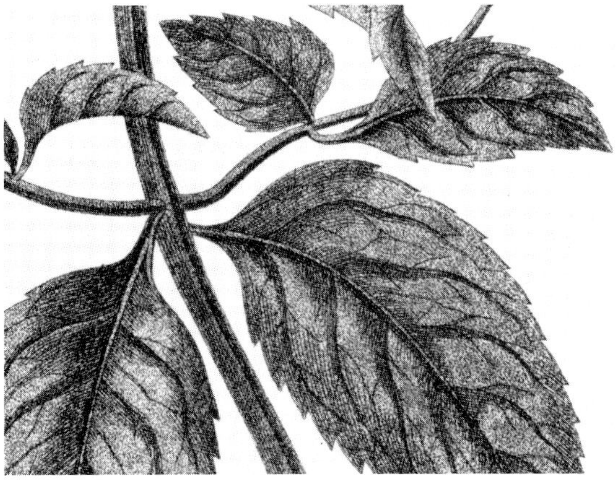

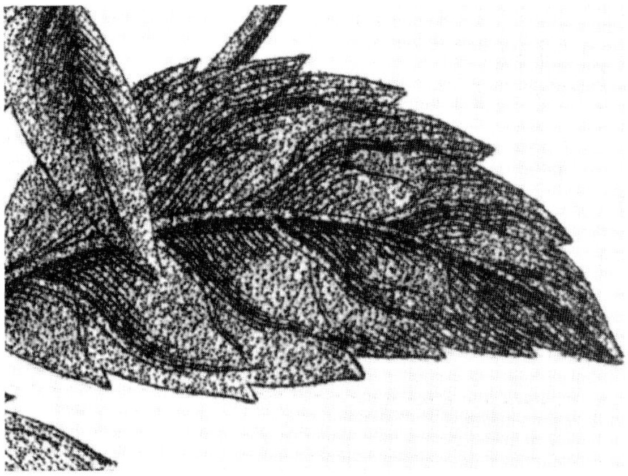

21. M: Piperita Vulgaris. Common Pepper-Mint. Line engraving, with overall stippled shading and some roulette shading, engraved by William Hibbert (English fl. ca.1760-1800) after a drawing by T. Sheppard (English fl. ca.1787). Forming a full-page illustration printed at Bath in 1797, plate printer unknown. Printed in black ink (platemark 26.5 x 19 cm.) on wove paper (top side, page size 33 x 22.5 cm.). In: W. Sole. *Menthae Britannicae.* Bath, printed by R. Cruttwell, 1798. Plate 8. Nissen BBI #1869.

A third example of botanical printmaking specifically designed not to be colored is found in the 24 illustrations to Sole's monograph of British mints. The author goes so far as to state in his preface that "as I have always been of the opinion that good plates are injured by coloring, I have endeavoured to procure such plates as need no coloring."

The illustrations are all "natural-size" etchings by William Hibbert after drawings by various artists, in this case the otherwise unknown T. Sheppard. His drawing has somewhat flattened the perspective of this decussate-leaved plant, perhaps because it was drawn from a dried specimen. Within an etched outline, Hibbert carefully indicated modelling with parallel-line shading, faithfully drew the venation, and showed close attention to detail in the flowering heads and the floral dissections. Finally, he laid a close-grained stipple over the whole image, covering shading and white spaces with equal persistence. The whole plate was then apparently etched with a single bite. It was a not wholly successful attempt to provide a range of monochrome tones to a line-drawn image. Since the tone is poorly modulated, it adds little to the perspective and says nothing about surface texture—the plant is actually glabrous but looks somewhat scabrous.

Certainly, no watercoloring could successfully compete with such dense grey tones. Unfortunately, it is this dense stippling that gives the plates the "rather dreary" quality that Blunt and others have noted. Nevertheless, the whole book is an ambitious production to have originated from such an unlikely town as Bath.

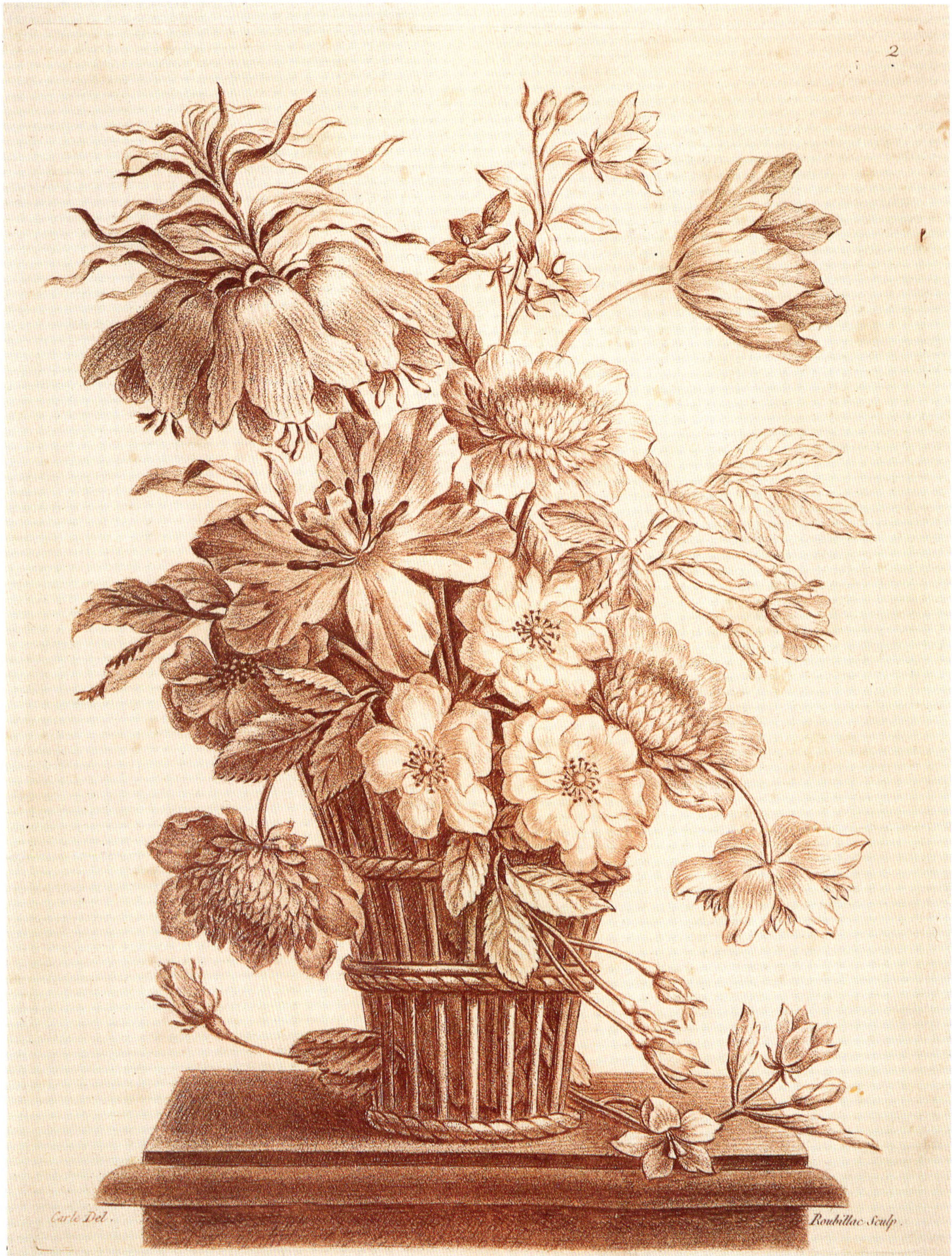
2
Carle Del.
Roubillac Sculp.

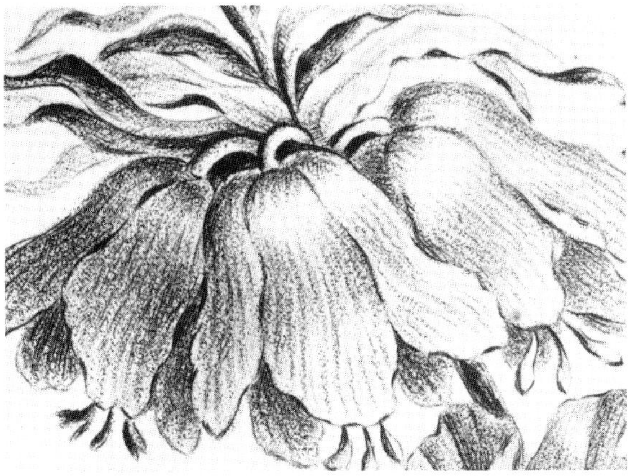

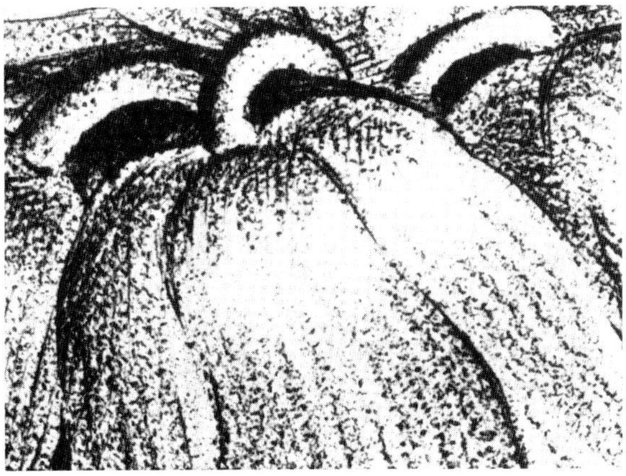

22. [Bouquet in a basket]. Crayon-manner etching, with some line-engraved details, engraved by [—] Roubillac (French 1739-ca.1773) after a drawing by [—] Carle (Swiss 18th cent.). Forming a full-page illustration printed at Paris in the 18th century, plate printer unknown. Printed in sanguine ink (platemark 27.7 x 21.2 cm.) on laid paper (wire side, sheet size 41 x 29 cm.). Probably ex: [—] Roubillac. *Etudes de Fleurs d'après Nature.* Paris, n.d. (In 23 cahiers of four plates each.) [Cahier ?], plate 2. Nissen BBI #1686.

Hibbert's not-too-successful attempt to give a finish resembling the shading of a soft pencil or chalk on rough paper probably reflected his lack of experience in using that effect. In England and France fine stipple engraving had long been practised with great success. Whereas Hibbert had added it as the final touch to an otherwise complete linear image, skilled stipple engravers strove to create the effect of the whole print having been executed in the "chalk manner," for which the visual syntax was entirely granular. This style of printmaking is another form of etching. Instead of drawing lines through the etching ground with a needle, chalk manner was achieved by dotting through the ground in closely laid stipple. Line was kept to a minimum to avoid hardening the effect.

In addition to using the etching needle or graver, the French masters developed some specialized tools for this manner of work, such as the double- or triple-pointed needle, the mattoir (like a little inverted spiny mushroom head) and the roulette (a mounted rotating wheel, usually about one-eighth inch wide, with the rim cross-cut to form lots of little pyramidal points). The latter was designed to create the effect of chalk strokes when rolled over an etching ground. Highly developed skills were needed to achieve a convincing imitation of chalk drawn over roughly textured paper. Additional skill lay in producing distinctions between the strokes that were intended to create perspective and modelling and those that were to represent surface textures of the subject.

Roubillac's bouquet intentionally aims at a loose, decorative style, rather than detailed botanical accuracy, but much more delicate effects could be obtained when required. There had long been a vogue for making drawings in red chalk or "sanguine," and the engraving shown here is a convincing attempt to simulate that style—no doubt Carle's original used that medium. There is some local line-engraved detail where sharpness or specially strong depth of color is required.

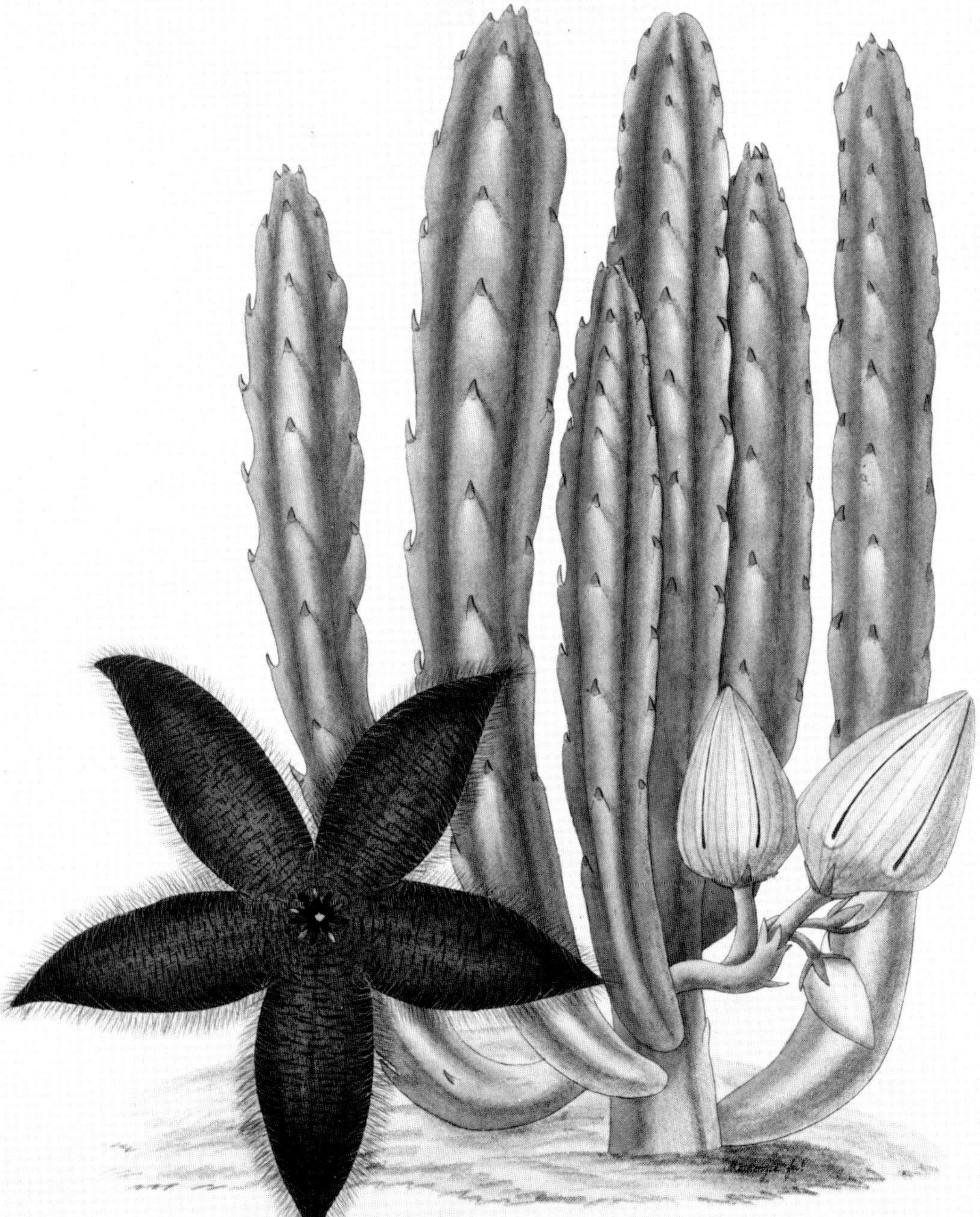

Stapelia ambigua

Publish'd as the Act directs, March 1.1797, by F. Masson.

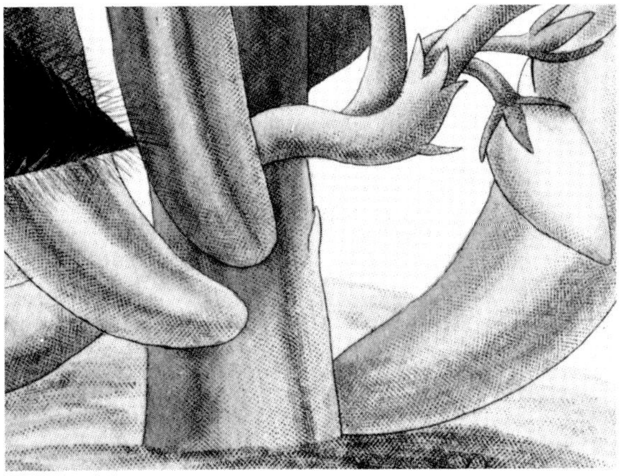

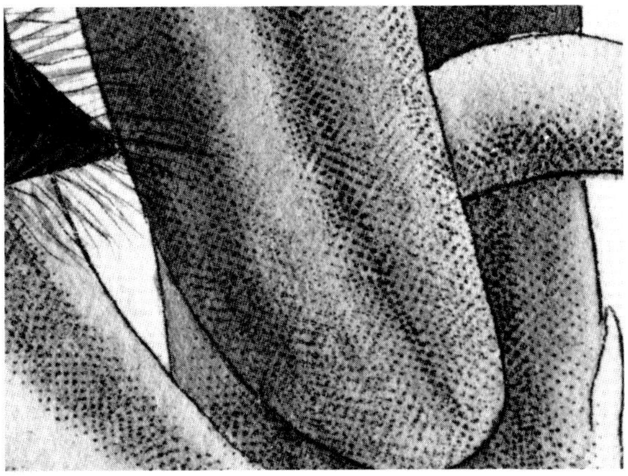

23. Stapelia ambigua. Line etching, with roulette shading, engraved by Daniel Mackenzie (English fl. ca.1784-1800) after a drawing probably by Francis Masson (Scottish 1741-1805). Forming a full-page illustration printed at London in 1797, plate printer unknown. Printed in black ink (platemark 35 x 25 cm.) on wove paper (top side, page size 36.2 x 26 cm.). Hand-finished with transparent and some opaque watercolor. In: F. Masson. *Stapeliae Novae*. London, G. Nicol, 1796[-1797]. Plate 12. Nissen BBI #1301.

The chalk effects obtained with the roulette could be utilized to provide local shading in a line etching, rather than using line shading or cross hatching. A great deal of the skill in using the roulette was to conceal any mechanical patterning that might be apparent if the teeth were cut in too regular a pattern. But roulette wheels varied. Some were, as described, broad, but others were very narrow with regular teeth and created a minute and evenly dotted line. The French specialists, in particular, developed many variations for special needs, but in England the roulette was not as widely employed.

The 41 plates in Masson's monograph of the succulent stapelias had great botanical interest in that they illustrated many species new to science. Furthermore, the illustrations were admired for their fidelity because, as Masson explained, "The figures were drawn in their native climate, and although they have little to boast in point of art, they possibly exhibit the natural appearance of the plants they represent, better than figures made from subjects growing in exotic houses can do." We do not know what these field sketches were like nor, indeed, whether it was Masson who drew most of them. They were possibly quite informally done. At any rate, Mackenzie, whose skills in engraving we have already seen in two previous exhibits (items #18, 19), translated the sketches into boldly etched outline drawings with a uniquely personal style of shading. He used the roulette to create patterned shadings that are strangely similar to the patent mechanical shading tints that Ben Day introduced to commercial art in the early 1880s.

Presumably both author and engraver were content with this oddly unchalklike style of roulette shading, though the living plant has no surface texture that would suggest this kind of treatment. The hand coloring includes some very carefully painted detail on the flower that is somewhat at variance with the freedom of the engraving.

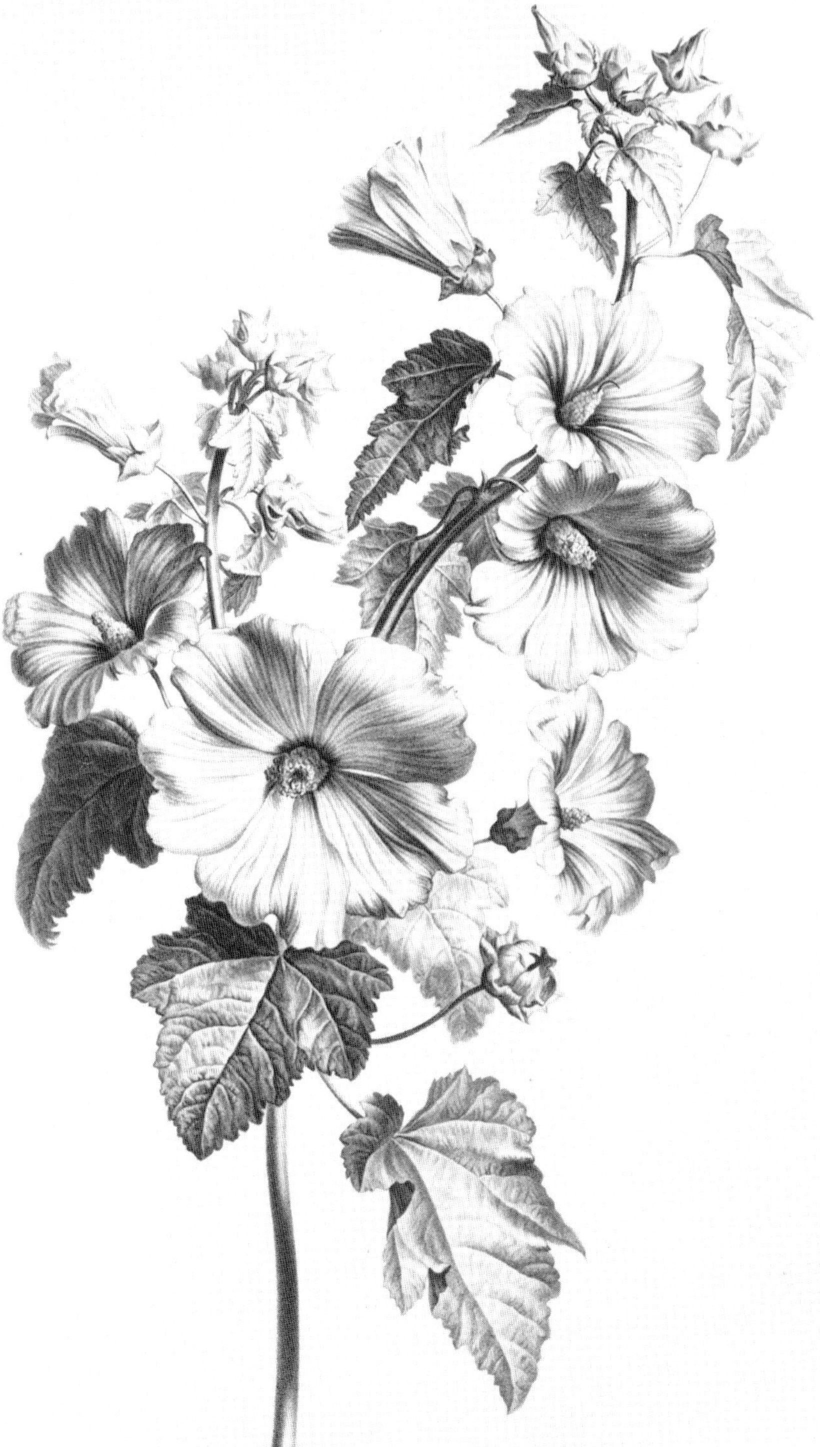

Lavatère à grandes fleurs.

Lavatera trimestris L.

G. Van Spaendonck del. L. C. Ruotte sculp.

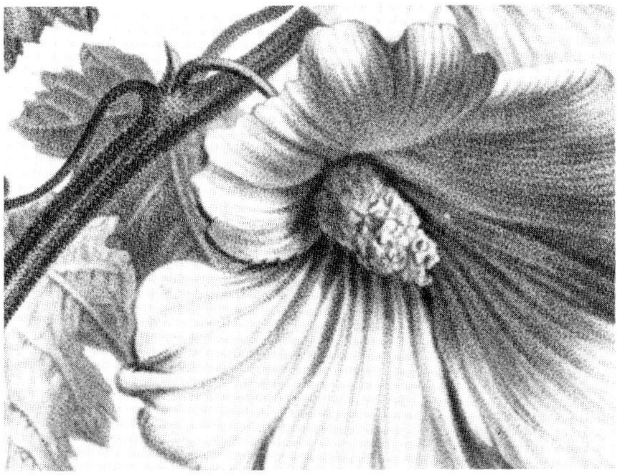

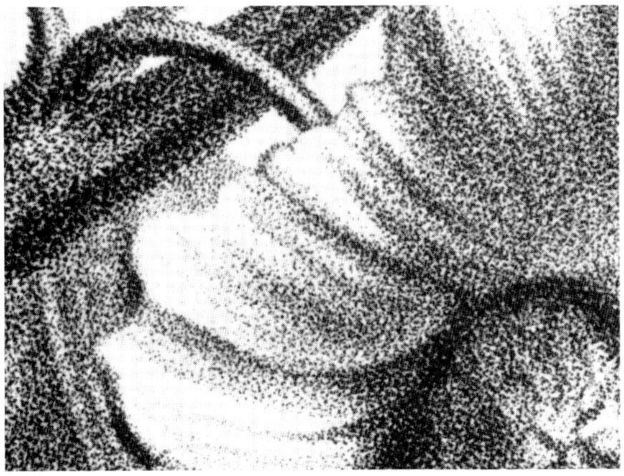

24. Lavatère a grandes fleurs / Lavatera trimestris. L. Stipple engraving, engraved by Louis Charles Ruotte (Père) (French 1754-ca.1806) after a drawing by Gerard van Spaendonck (Dutch 1746-1822). Forming a full-page illustration printed at Paris in 1801, plate printer unknown. Printed in black ink (no platemark visible) on laid paper (top side, watermark "T. Dupvy fyn," sheet size 49 x 33.5 cm.). Ex: G. van Spaendonck. *Fleurs Dessinées d'après Nature.* Paris, chez l'Auteur, [1801]. Plate [5]. Nissen BBI #1879.

By the turn of the 18th century, stipple engraving in both England and France had achieved a high level of popularity and its practitioners were producing work of unrivalled skill and artistry. However, it failed to gain much acceptance for botanical work in England, where etching and line engraving held the stage, whereas in France it was especially favored for plant illustration.

We saw how coarse chalk strokes could be successfully imitated with crayon-manner engraving (item #22). But stipple could be used in a much finer-grained manner that was capable of indicating the most refined nuances of light and shade, of form and texture. It was ideal for delineating plant surfaces and structures; only when delineating the tiniest structures, such as plant hairs, did it prove unsuitable.

The example shown here is one of a series of 24 prints that exist in plain, hand-colored (under Spaendonck's direction) and color-printed versions. This monochrome version enables us to appreciate the delicate subtleties of pure stipple technique without the seductive distraction of color. It has been said that those skilled in the peculiar manner of "pecking" the grounded copper—using a slightly down-curved burin—were capable of working at great speed. Ruotte was certainly one of the more gifted stipple engravers of his day, and the prints he made after Spaendonck's delicate drawings have moved such authorities as Blunt, Stearn and Dunthorne to rate them among the finest engravings of flowers ever made.

Although this style of engraving was normally dotted through an acid-resisting ground and then bitten with mordant, it was possible to peck the plate sufficiently firmly to create incisions that would print without etching. Under high magnification we can see that some of Ruotte's stippling has the clean triangular shape of the burin's tip, indicating the latter manner of working. With careful control it was possible to use this manner to complete the more delicate passages in a stipple engraving.

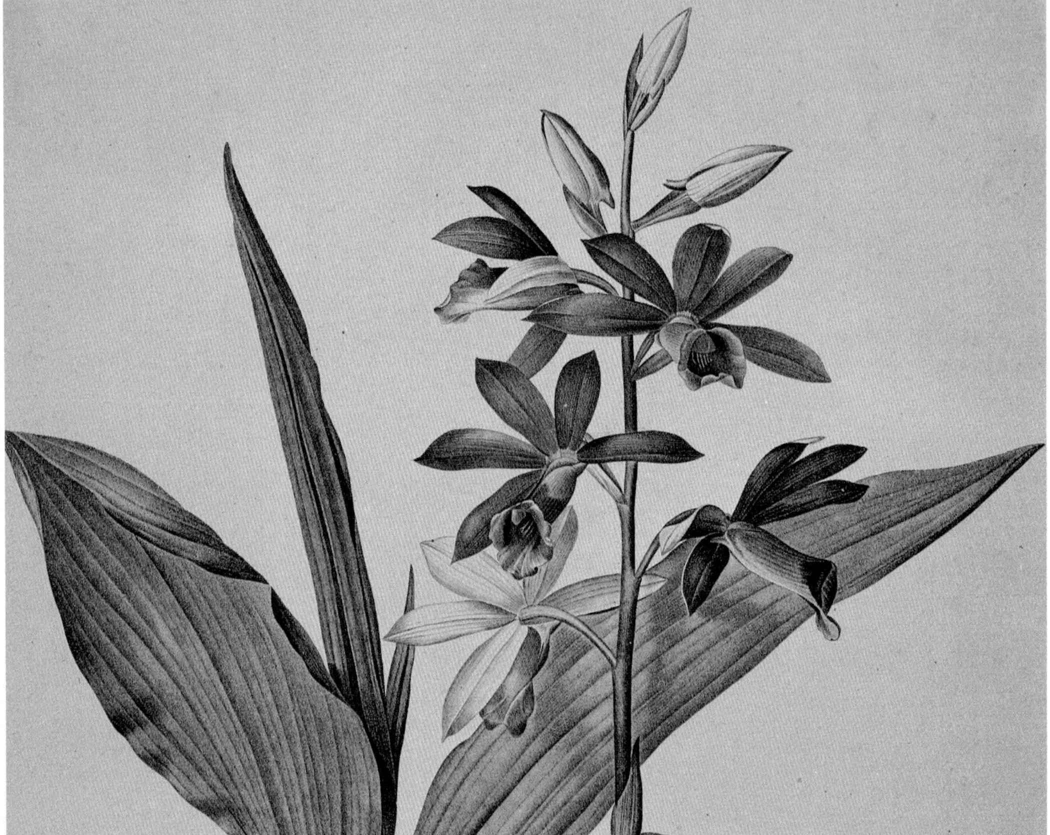

Limodorum Tankervillæ *Limodore de Tankervill*

P. J. Redouté pinx. de Gouy sculp.

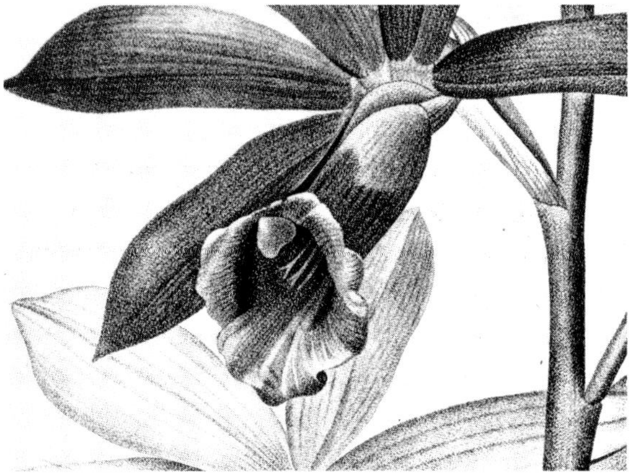

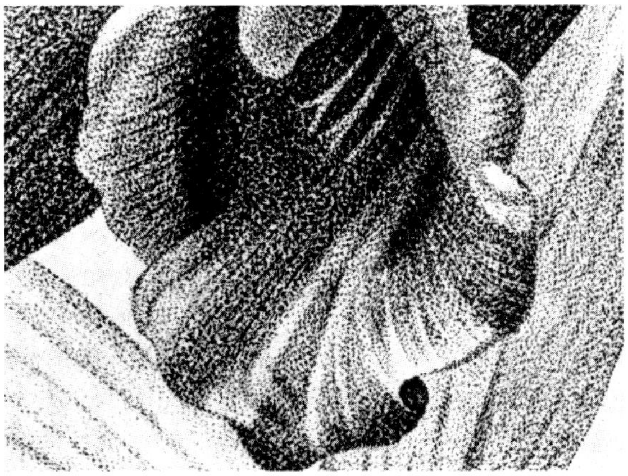

25. Limodorum Tankerwillae. Limodore de Tankerwill. Stipple etching, with some roulette tints, engraved by [—] de Gouy (French fl. ca.1800-1820) after a drawing by Pierre-Joseph Redouté (Belgian 1759-1840). Forming a full-page illustration printed at Paris in 1803, plate printer unknown. Printed in black ink (no platemark visible) on ochre-tinted wove paper (top side, sheet size 52.3 x 35.3 cm.). Ex: P.-J. Redouté. *Les Liliacées.* Paris, chez l'Auteur, 1802-1816. Plate 43. Nissen BBI #1597.

Undoubtedly the greatest exponent of botanical stipple engraving was P.-J. Redouté. He had first seen the potential of the medium as used by London printmakers whilst visiting there to work on the plates for L'Héritier's Sertum Anglicum *(item #20). His originals for that work, done in grey tonal washes, were prepared for line-engraved plates but they could have served equally well as patterns for stipple engraving.*

We are told that Redouté spent several years developing his art with the help of skilled French engravers, and in particular perfected its use in conjunction with à-la-poupée color printing. Stipple engraving, as practised in England, was not regarded as a very demanding craft. The eminent engraver Sir Robert Strange protested that "From the nature of the operation and the extreme facility with which it is executed, it has got into the hands of every boy, of every print seller in town, of every manufacturer of prints, however ignorant and unskillful... The art is in itself extremely limited, admits of little variety, and is susceptible of no improvement." Redouté's engravers would have little difficulty in mastering the basic technique. In fact, they improved its range of expression by skillfully combining it with roulette work. We must assume that the years of experimentation that Redouté undertook were largely directed at refining the accurate stippled depiction of plant structures and textures, and most especially at perfecting the production of correct tonal values for printing in color.

It was discovered by English printers that stipple engravings printed most successfully from plates that had been well used. A number of black impressions were run off to take the sharpness off the plate. Redouté's printers also took some black impressions from plates for both the Liliacées *and the* Roses. *For the interest of connoisseurs Redouté included a set of black plates as a parallel series to the usual color-printed versions in special issues of both books.*

Significantly, the black impressions are always printed on paper with a strong ochre-yellow tint, like the example shown here. Since black has a much greater force than the delicate colored inks washed with thin watercolor that Redouté normally used, black impressions on reflective white paper would have produced prints with grossly exaggerated tonal contrasts. By using paper devoid of brilliance, he was able to subdue that contrast and produce black prints that enabled the reader to appreciate the purity of his engravers' stipple and roulette technique.

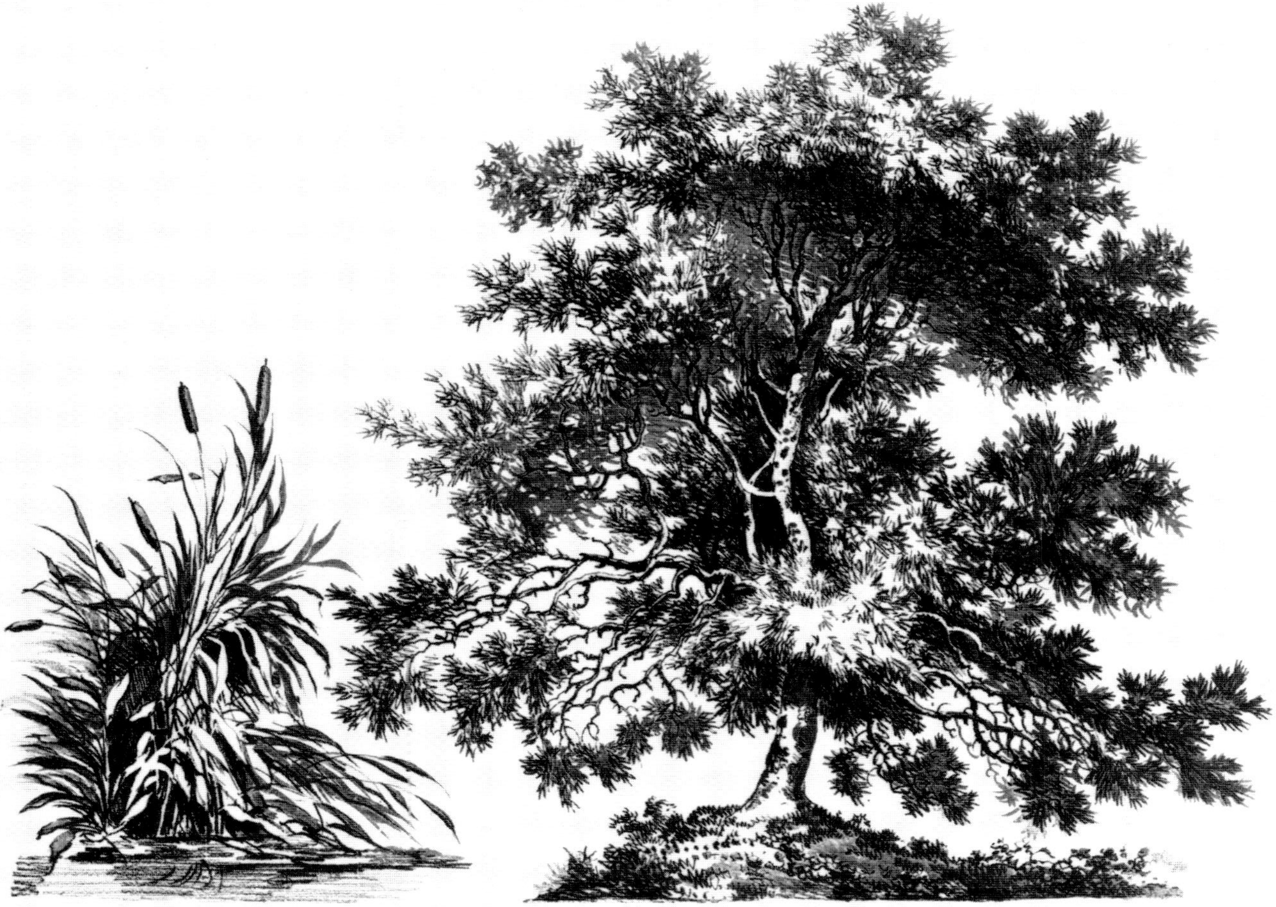

J. La Porte del | Bull-rush | Occidental Plane | et Sculp.

London Publish'd Decr. 7 1796 by Thos. Macklin, Poets Gallery, Fleet Street.

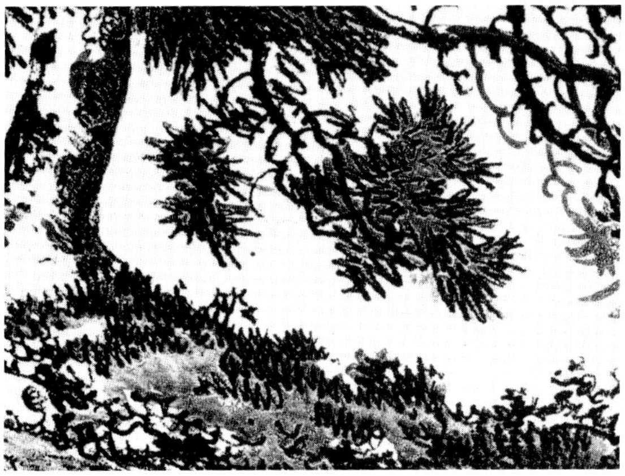

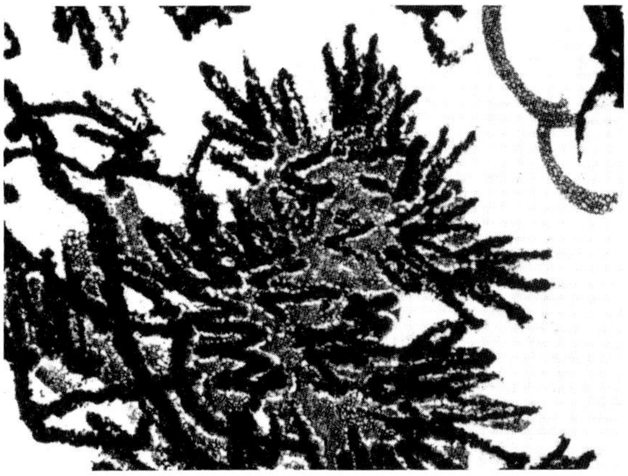

26. Bullrush *and* Occidental Plane. Soft-ground etching and lift-ground aquatint, engraved by John Laporte (English 1761-1839) after his own drawing. Forming a full-page illustration printed at London in 1796, plate printer unknown. Printed in black ink (platemark 37.5 x 48 cm.) on wove paper (top side, watermark "1794 J. Whatman," page size 38.3 x 52 cm.). In: J. Laporte. *Characters of Trees.* [London], Macklin, [1795-1801]. Number 5, plate [2]. Nissen BBI #1142. Lent anonymously.

The delineation of plants without the hard edge of outline drawing was possible with line or stipple engraving, as demonstrated by Bloemaert and Ruotte (items #15, 24). The hardness of etched outlines could be subdued by vigorous plate wiping, as was done for Haid's print of Pancratium *(item #10). However, the least linear manner was "soft-ground" etching. This example is one of several tree studies engraved by Laporte with this technique.*

The unengraved copper was covered with a soft form of etching ground and then a sheet of clean paper was laid over it. The artist made either an original drawing or traced an existing design on the paper with firm pressure on the pencil. When the paper was lifted off the surface it pulled some of the wax away from the copper wherever the pencil had pressed. The exposed, uneven or "crumbly" line was then bitten with a mordant in the usual way, producing a broad, soft-edged line. For sketches of plant habit and foliage it produced a print with a chalk-like quality. Soft-ground etching can bear a close similarity to chalk-style lithography.

Apart from its use for spontaneous sketching, soft-ground etching had little application in botanical work. Laporte, an English landscape painter, drawing master and etcher, used the process for numerous prints illustrating his ideas and recommendations for foliage, tree and landscape drawing. The characteristics of the process are quite apparent in this example, but it should be noted that Laporte also added some background shading in aquatint.

This print is from the series of Laporte's tree studies that Henrey lists as having been published between 1798 and 1801. However, this example and seven others from an incomplete series of 12 bear copyright dates of either January 1st or December 7th 1796.

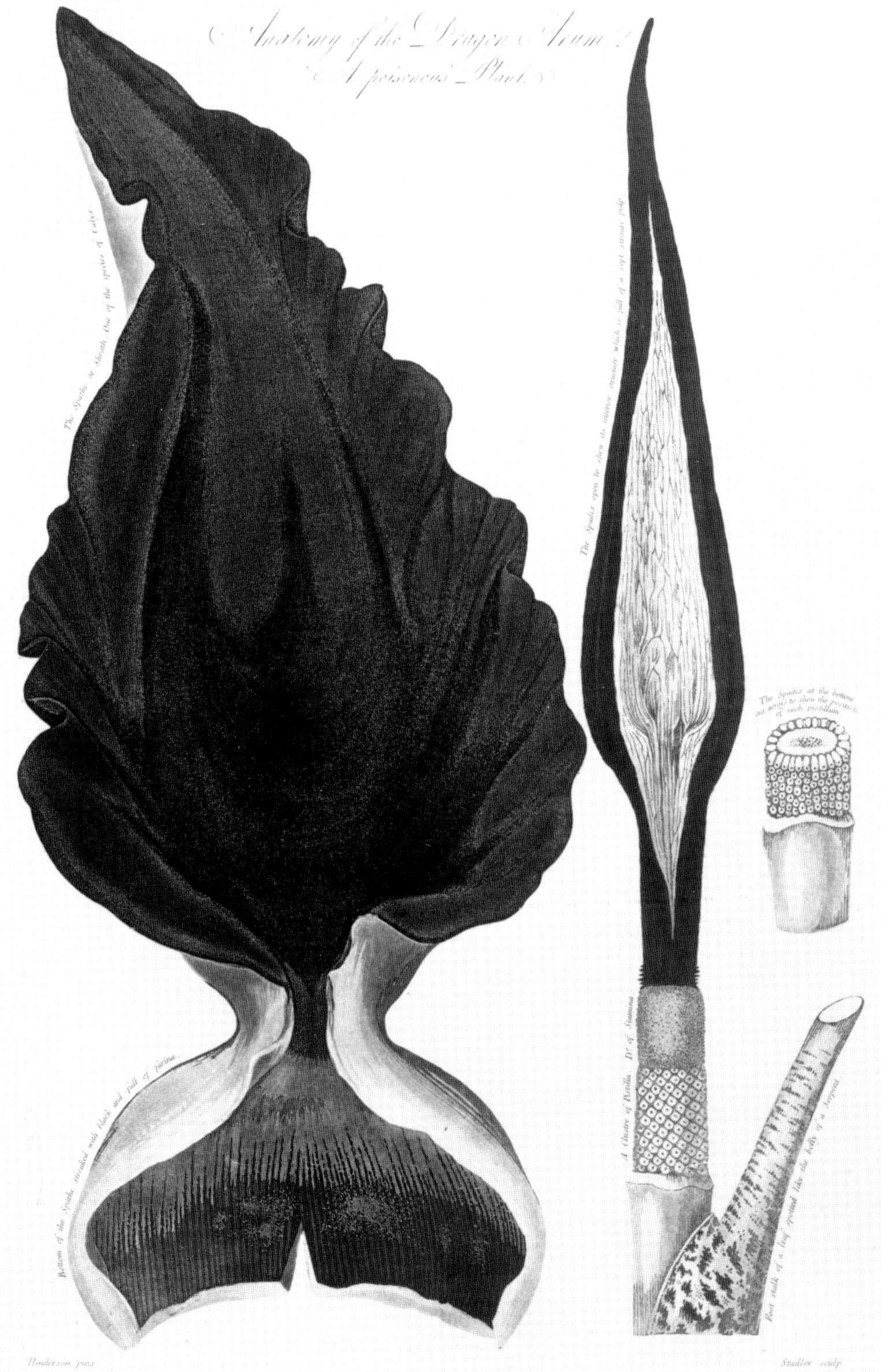
Anatomy of the Dragon Arum?
A poisonous Plant.
The Spathe or Sheath. One of the species of Calyx.
The Spadix open to shew its interior structure which is full of a soft stringy pulp.
The Spadix at the bottom cut across to shew the position of each pistillum
A Clustre of Pistilla. D° of Stamina
Bottom of the Spathe streaked with black and full of farina.
Foot stalk of a leaf spotted like the belly of a Serpent.
Henderson pinx
Stadler sculp
London Published by Dr Thornton

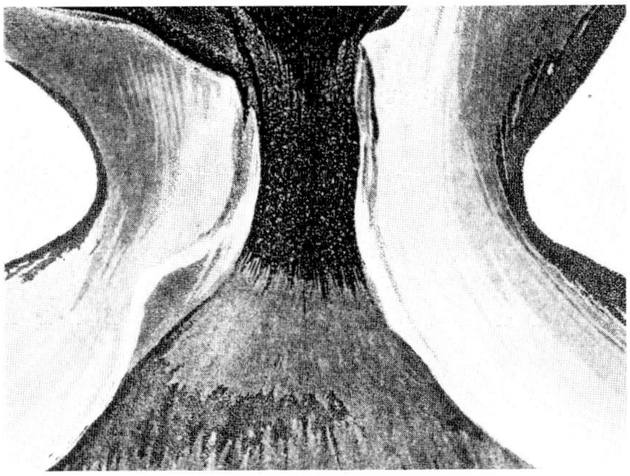

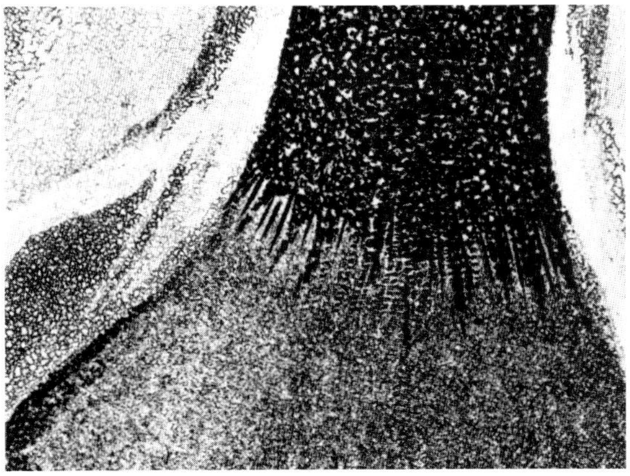

27. Anatomy of the Dragon Arum / A poisonous plant. Spirit-ground aquatint, with some lift-ground etching, roulette and line engraving, engraved by Joseph Constantine Stadler (German fl. ca.1780-1815) after a drawing by Peter Charles Henderson (English fl. ca.1791-1829). Forming a full-page illustration printed at London in 1802, plate printer unknown. Printed in black ink (platemark 49.3 x 33 cm.) on wove paper (top side, watermark "HS 1809," page size 58.5 x 46 cm.). In: R. J. Thornton. *New Illustration of the Sexual System of...Linnaeus.* London, [1799-] 1807. Plate [?]. Nissen BBI #1955.

We saw, in passing, that Laporte used aquatint to supplement the tonal limitations of soft-ground etching. Aquatint—that is, a tint created by aqua-fortis*—was a purely tonal form of "wash" etching. As such, it scarcely enabled a printmaker to "draw" plant structures as any linear process permitted. Aquatint was invented in France in the mid-18th century and its most successful use was for broad-toned landscape work. However, it was rarely used as a pure medium and was almost always supplemented by some line etching. Generally speaking, it was a speedy medium and thousands of aquatints were produced in England after its introduction there in 1775 until the 1820s, when the process was largely displaced by chalk-style lithography.*

A fresh copper plate was covered with a fine scatter of powdered resin—so finely spread as to leave spaces of bare metal exposed between the particles. Heating the plate greatly from below caused the particles to melt and spread a little, many fusing with their neighbors, but still leaving a reticulation of tiny spaces of bare copper. Alternatively, a fresh copper could be flooded with resin dissolved in a spirit. When the spirit evaporated the resin precipitated in a dense even pattern of tiny islets. In either method the exposed copper was then etched and the resin removed. The result was a fine intaglio reticulation that printed as a flat grey tint. However, an aquatint image was normally created from several bitings before removal of the resin. The artist had to "stop out" selected areas of the first bite with acid-resisting wax and re-etch for a darker tint. By repeating the steps he built up a tinted picture composed of increasingly darker layers of flat tints.

Flowers are generally rather complicated subjects for the tonal syntax of aquatint, but the method was successfully used for illustrating fruits. This example is from one of the few botanical works that used the medium. It is a spirit-ground aquatint with the most skillful use of progressive etchings and stopping-out. Apart from the center of the spathe, which is strengthened with roulette and a few strokes of the burin, every detail is aquatinted, even what appear to be lines. Judged by any standard it is a uniquely dramatic and unusual print, but the choice of heavily bitten aquatint to successfully obtain such powerful tones is a particularly daring stroke. Stadler, more at home doing landscape aquatints, created a plant portrait the like of which can scarcely be found elsewhere in botanical printmaking.

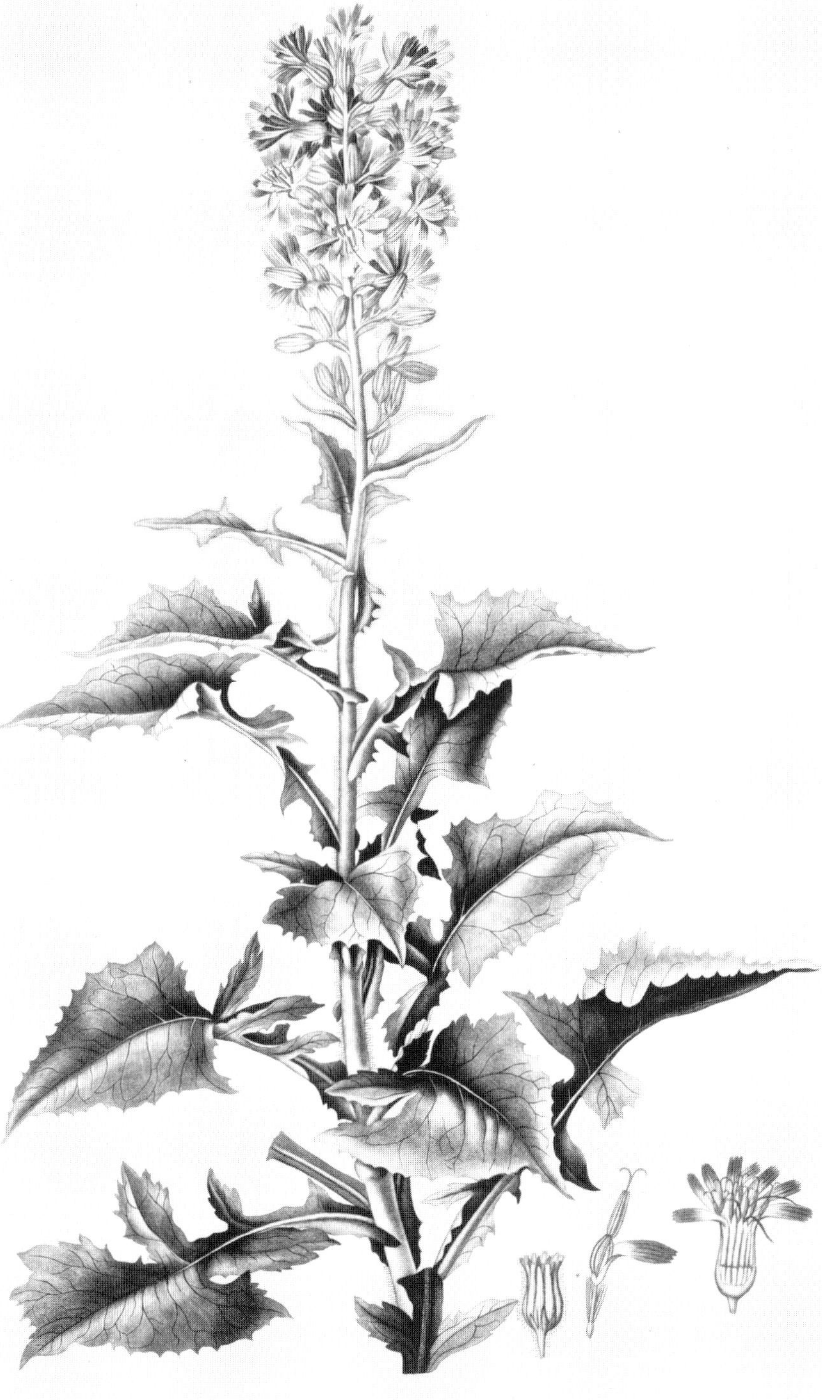

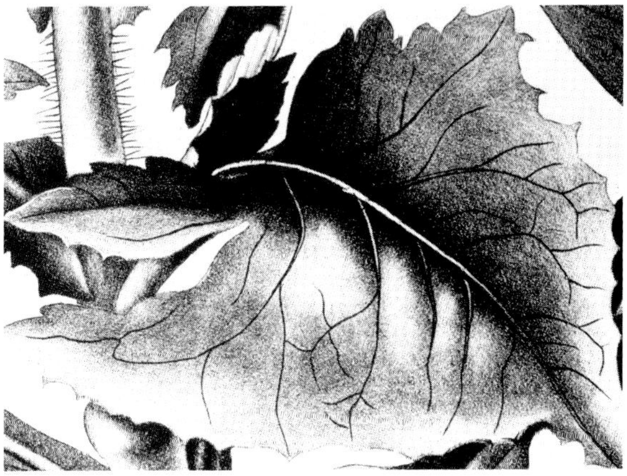

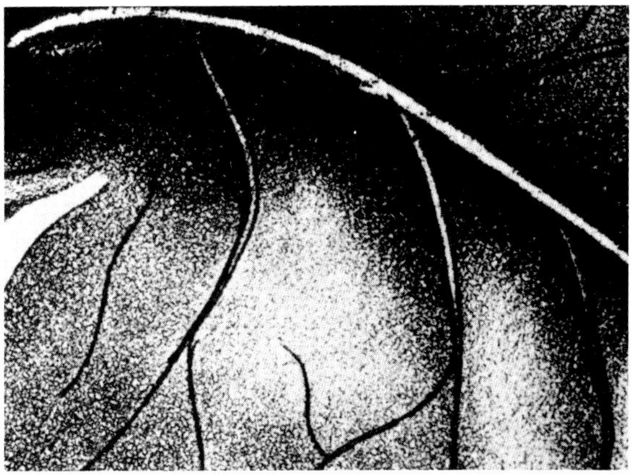

28. [Cicerbita alpina #78] 2. (reversed). Line engraving and etching, with dust-ground aquatint, engraved by Peter Beckenham (?Austrian 18th cent.) after a drawing by Johann Knapp (Austrian 1778-1811). Forming a full-page illustration probably printed at Vienna ca.1806, plate printer unknown. Printed in black ink (platemark 50.5 x 37.5 cm.) on wove paper (top side, watermark "I. C de R. IM-HOF, KLR. REAL.," sheet size 62.5 x 46 cm.). Prepared for: [Johann, Erzherzog von Österreich. Unpublished Icones Plantarum Austriacarum Ineditae. Vienna, ca.1806 +].

The process of stopping-out and re-biting an aquatint was apt to produce rather abrupt "steps" between tints and, at its simplest, created an effect akin to the layered contour shading of a relief map. Aquatinters sometimes used as many as 12 or more separate bitings to avoid this effect. But an alternative solution existed. Instead of total immersion in acid for each bite, small pools of the mordant could be applied to parts of the plate and spread around with a brush, a process known in England as "feathering." All areas to be tinted were usually given a very light, finely grained preliminary tint before feathering began. A small puddle of mordant was then applied first to a passage required to be deeply bitten and then spread gradually to include lighter areas before being washed off. The printmaker was, in effect, "painting" on the plate and could produce smooth tonal gradients ranging from nearly solid black to pale silvery greys. It required a rare degree of skill to achieve fine results, but in clever hands it was possible to create prints of a character that could scarcely be approached by any intaglio process we have seen, perhaps excepting stipple engraving.

The print shown here is a remarkable example of the skillful application of this method to an outline etching of a plant. Virtually all the light and shade, modelling, and surface textures are created by aquatinting of a subtlety that foreshadows what later 19th-century photomechanical halftone would achieve. This stately portrayal of a humble weed was a laborious engraving task, making the process quite uneconomic for normal botanical illustration.

It is the work of the otherwise unknown Peter Beckenham and is from a set of 42 in the Institute's collection—one of only five known sets in the world. These handsome portraits were commissioned by Archduke Johann of Austria, who apparently planned to publish a magnificent iconography of alpine plants based on his private collections. The project obviously foundered, for reasons that are not recorded, and all that remains are the scattered sets of these prints, each different in content. A collection of 185 watercolor originals by Knapp, out of the 300 reputedly made for this work, is preserved at the Institut für Systematische Botanik in Graz. No trace of a printed text or title page has yet been found.

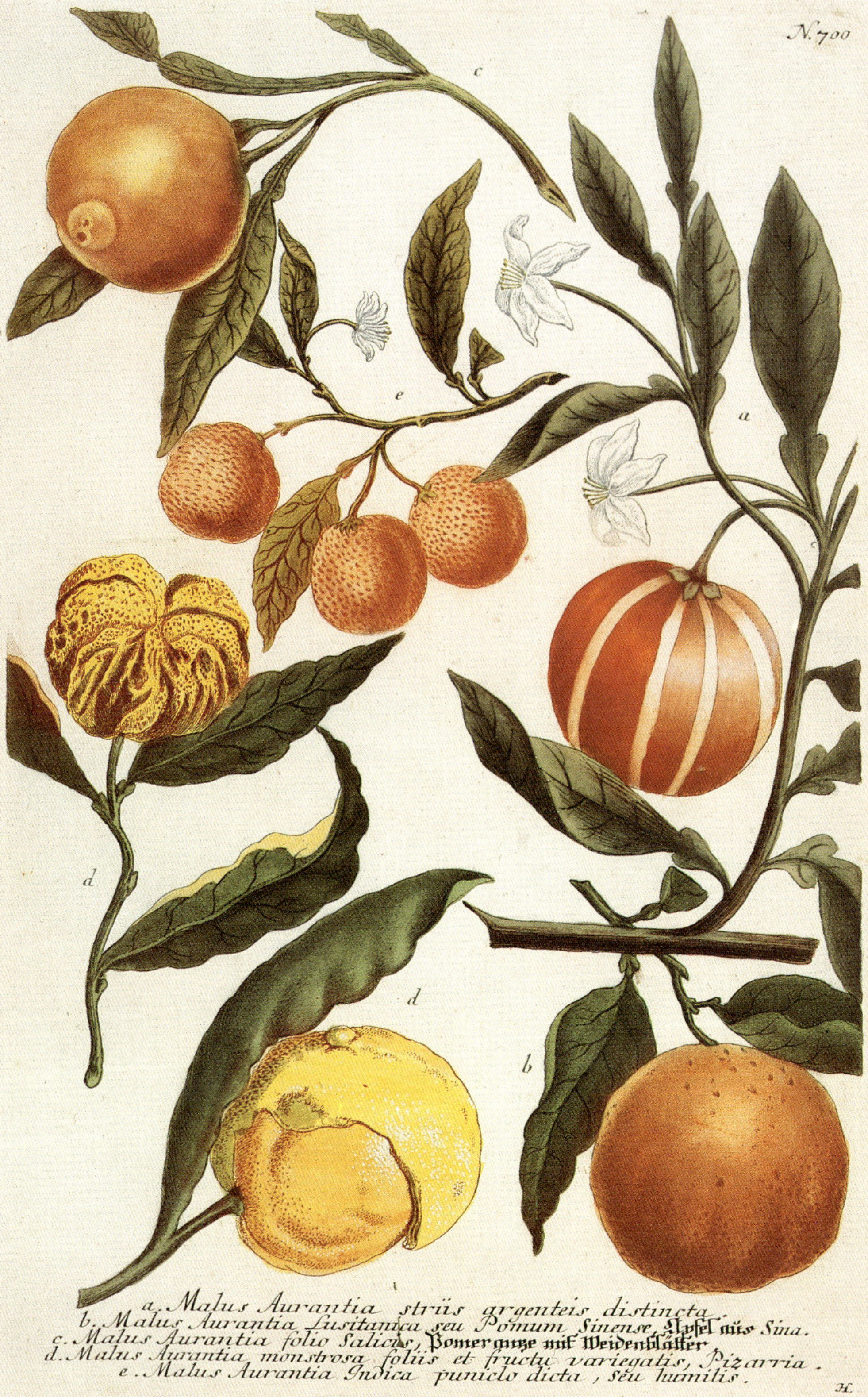

a. Malus Aurantia striis argenteis distincta.
b. Malus Aurantia Lusitanica seu Pomum Sinense, Apfel aus Sina.
c. Malus Aurantia folio Salicis, Pomeranze mit Weidenblätter.
d. Malus Aurantia monstrosa foliis et fructu variegatis, Pizarria.
e. Malus Aurantia Indica puniclo dicta, seu humilis.

H.

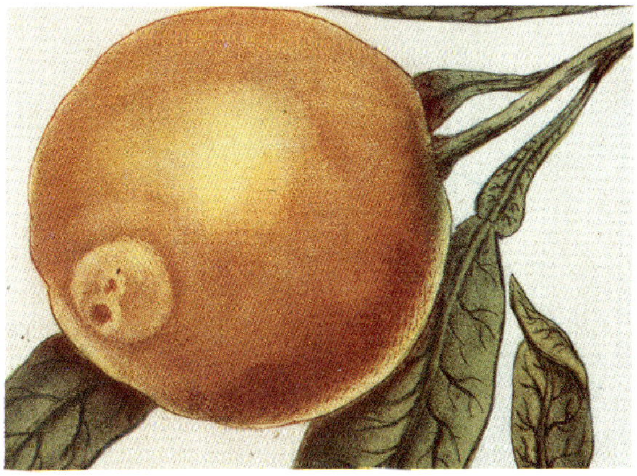

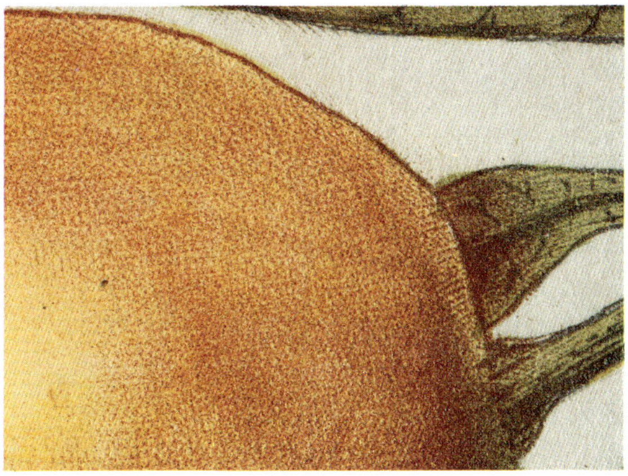

29. a. Malus Aurantia strüs argenteis distincta... Line etching with line shading and rocked tints, engraved by Johann Jakob Haid (German 1703-1767), probably after a drawing by Georg Dionysius Ehret (German 1708-1770). Forming a full-page illustration printed at Ratisbon in 1742 by J. J. Haid. Color-printed, à la poupée, in black, green and russet inks (platemark 33.5 x 21.5 cm.) on laid paper (top side, sheet size 39 x 24.5 cm.). Hand-finished with transparent and some opaque watercolor. Ex: J. W. Weinmann. *Phytanthoza Iconographia.* Augsburg, B. Seuter, J. E. Ridinger & J. J. Haid. (Regensburg, Lentz), [1735-] 1737-1745. Plate 700. Nissen BBI #2126.

Some 60 years before Beckenham engraved his elegant aquatint-toned print, there appeared a large series of illustrations that aimed at a similar, if less successfully achieved, effect using rather different means. These too were etched in outline and shaded with smoothly modulated tones, but the tones were mechanically produced by the mezzotint rocker.

This device (aptly called berceau *in French) may be likened to a broad-bladed chisel, with the cutting edge ground into a curved, rocker-like, shape. The flat side of the chisel is evenly cut with many fine, parallel longitudinal grooves. The other, bevelled, side thus terminates in a finely toothed cutting edge. When this curved serration is firmly rocked back and forth on a polished copper plate, it creates a row of tiny pits with the metal forced up into a sharp rim around each. Repeated rocking over an area will produce innumerable pits. If the rocker is turned to different angles and the area repeatedly cross-rocked, the density of the pits completely eliminates all trace of polish from the original surface. Rocking may be done in local areas or over an entire plate, and if the burred texture is inked over, it will print as a solid velvety-black tint.*

Mezzotint means, literally, "middle-tint" and to achieve that value the fully textured copper is firmly rubbed with a blunt-bladed scraper; with persistence, this can restore a smooth brightness to the surface that will print as white. Between the two extremes it is possible to achieve an infinitely varied gradient of tone and, since scraping can be done over broad or minute areas, very sophisticated tonal images may be created.

Haid added tones to this print of citrus fruits by applying local rocked tints and modulating them with the scraper. This is one of over a thousand similar illustrations, all done with more speed than grace. It was printed in colored inks before being heightened with somewhat clumsy hand coloring. Haid was capable of much more refined work, as his print of Pancratium *(item #10) demonstrates, and with care could have achieved much more convincing modelling on this print. But the author, Weinmann, could be a mean and unrewarding employer, as Ehret had discovered, and we may assume that Haid did no more than his remuneration required.*

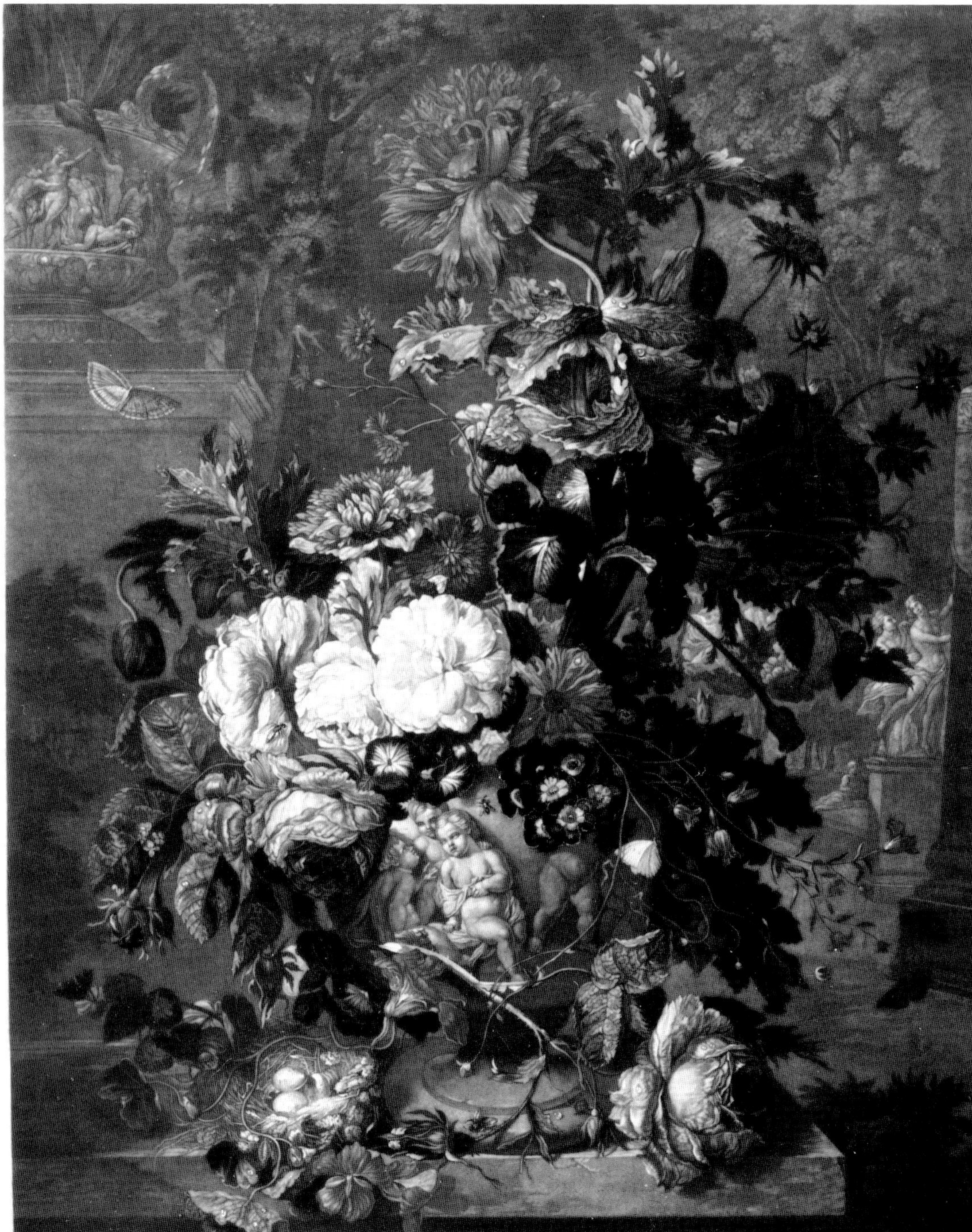

A FLOWER PIECE.

In the Cabinet at Houghton.

Richd. Earlom Sculpsit

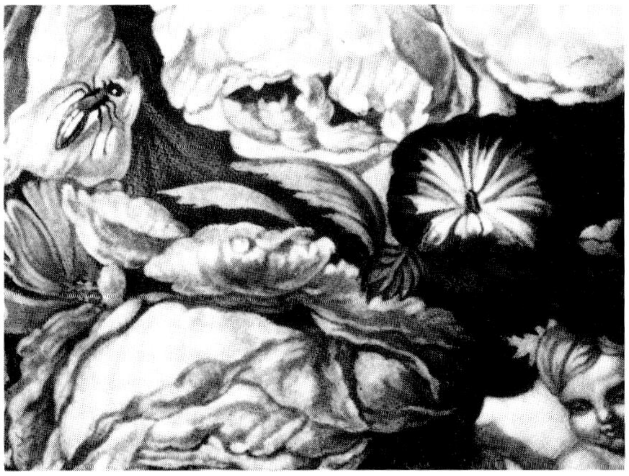

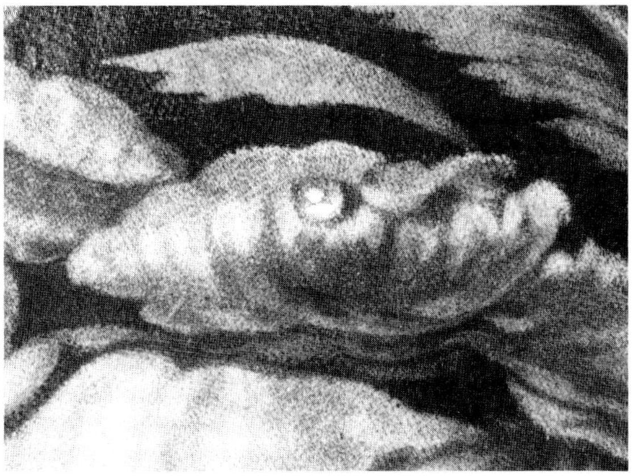

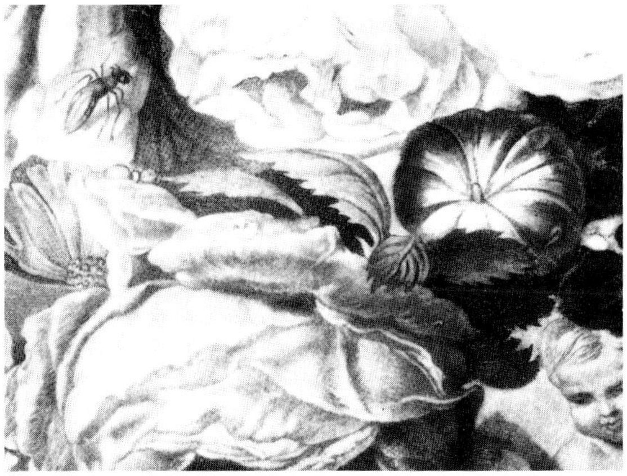

30. A FLOWER PIECE / In the Cabinet at Houghton. Mezzotint engraving, with some line, stipple and roulette details, extensively reworked, engraved by Richard Earlom (English 1743-1822) after a painting by Jan van Huysum (Dutch 1682-1749) redrawn by Joseph Farington (English 1747-1821). A separately published picture printed at London, plate printer unknown, and issued by John Boydell in 1778. Printed in black ink (engraved area 50.5 x 39.5 cm.) on wove paper (top side, sheet size 55 x 41.7 cm.). Hand-finished with transparent and some opaque watercolor.

Mezzotint was first discovered in the mid-17th century, and Prince Rupert of Bohemia must receive the greatest credit for its development. The earliest excursions into the medium treated the rocked tint as an additive process, just as Haid did for his citrus print (item #29). Mezzotint emerged as a wholly individual medium when the Prince and his assistant successfully developed the method of laying solid grounds and, in the process of doing so, invented the rocker. Having achieved this, he was able to demonstrate mezzotint's true potential as a subtractive process, for all images are created by scraping lights out of the solid black tint. Its full tonal capabilities were, from the outset, used for reproducing copies of oil paintings. Earlom's skillfully scraped mezzotint fits the classic mould. In rich velvety tones he created a monochrome translation of Jan van Huysum's oil painting.

Since the mezzotint ground could be scraped into such delicately modulated tones, it is not surprising that the repeated plate wiping that precedes printing, coupled with the immense force of the printing press, should weaken this texture. Mezzotint plates soon wore out and, after about 70 impressions had been taken, the plate had lost much of its richness and contrast. If further prints were required, the ground had to be rejuvenated. This impression of Earlom's print has had extensive stipple and roulette reworking to restore weakened passages. The Institute also has an earlier and richer impression taken from the same copper (illustrated in Flora Portrayed: Classics of Botanical Art from the Hunt Institute Collection, *1985). Minute comparison of the two reveals a surprising amount of alteration in many details, as illustrated here. The original freshness of pure mezzotint scraping could never be regained by adding lines or dots to weakened tones for this changed the character, as well as the detail, of the print. In an effort to conceal these tamperings and to compensate for an overall lightening in tone, the printer could also over-ink later impressions—as he did here. The earliest impressions of this print were uncolored and it is likely that coloring was used as another means of concealing the reworkings.*

This constitutional weakness of mezzotint militated against its successful use in book illustration except for extremely limited editions. This print was obviously "extended" to meet a greater demand than the original plate would satisfactorily allow, and copies are therefore found in various states of wear and rejuvenation.

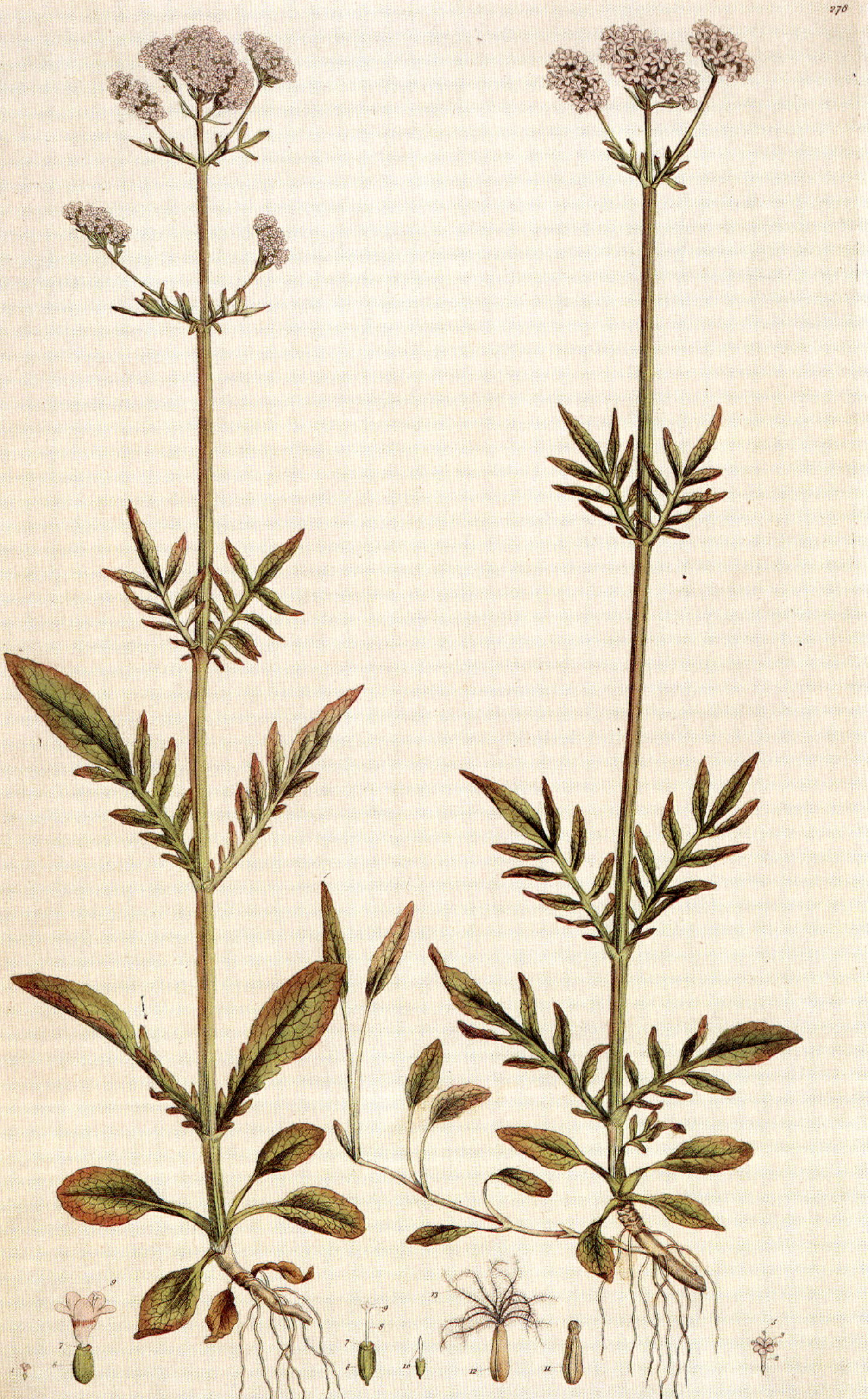

Valeriana dioica.

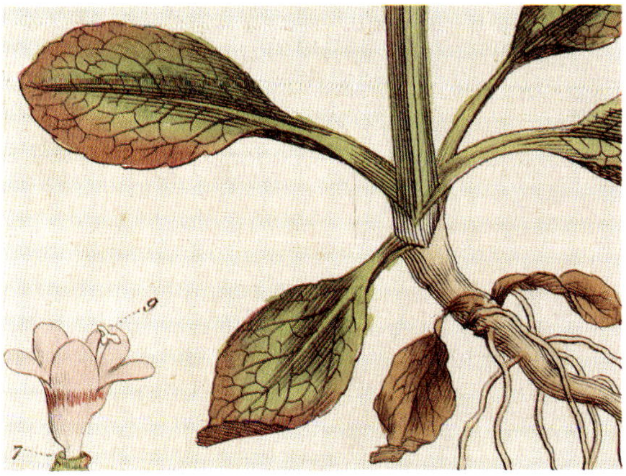

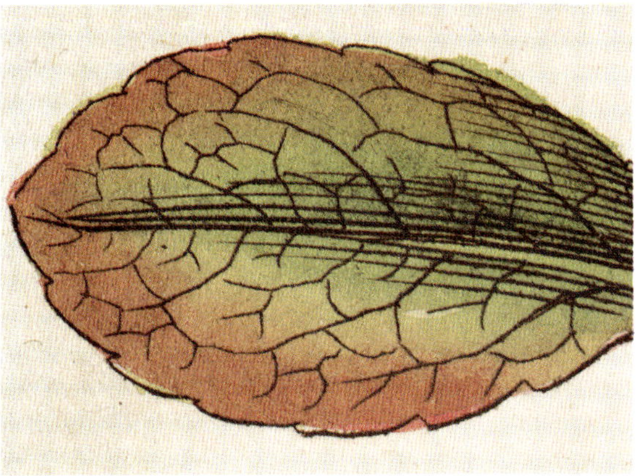

31. Valeriana dioica. Line engraving, outline style, with some etching, engraved after a drawing probably by Sydenham Teast Edwards (Welsh 1768-1819), engraver unknown. Forming a full-page illustration printed at London in 1783, plate printer unknown. Printed in black ink (platemark 43 x 26 cm.) on laid paper (top side, watermark "IV," sheet size 46.5 x 28.3 cm.). Hand-finished with transparent watercolor. Ex: W. Curtis. *Flora Londinensis.* London, The Author, [1775-]1777-1798. Plate 278. Nissen BBI #439.

The splendid plates in Curtis's Flora Londinensis *have been justly admired by botanists and collectors. The folio format was chosen to permit life-sized depiction of most species but the generous dimensions of the plates also had the effect of giving a special dignity to even the lowliest wayside flowers. The quality of the engraving is unremarkable, and much of the visual appeal can be attributed to attractive hand coloring, all the subjects being represented in their "true colors."*

Hand coloring could add a new dimension to the botanical information in a plant portrait, quite apart from enhancing aesthetic appeal. However, as a hand process it was always subject to individual variation, either because a single colorist was inconsistent or because more than one colorist was employed to paint the run of a print. Furthermore, pigments were not always consistent in quality, appropriate ones were not always available, and certain white and red ones tended to turn black in the sulphurous air of industrial cities.

The normal procedure was for the artist to prepare a colored "pattern" for each plate as a guide to the colorist and, we assume, to inspect the latter's work before the prints were released for publication. Without authoritative supervision, hand coloring was of dubious botanical value.

Curtis is known to have employed some 30 colorists for the 432 plates in the Flora, *which is reported to have run to an edition of about 300 copies. Publication commenced in 1775 and took 23 years to complete. Unfortunately, in the early stages at least, Curtis did not adequately supervise his colorists or check the accuracy of their work. One customer, a Dr. William Hird, was moved to write that "...my friend Curtis will give me leave to tell him that the colouring of every plant in the first number in my possession is by no means equal to the specimen Cousin Nancy Freeman left with me."*

The Flora *appeared "...periodically in Numbers price 5s. coloured, or 2s. 6d. plain," but we are also told that "some copies finished with extraordinary care were sold at 7s. 6d." Although each print represented a faithful reproduction of the engraver's original work, the coloring ranged in quality from that of Dr. Hird's unacceptable copy to the extraordinary finish of the expensive special issue.*

This example is typical of the kind of transparent coloring found on thousands of botanical prints produced all over Europe in the later 18th and throughout the 19th centuries.

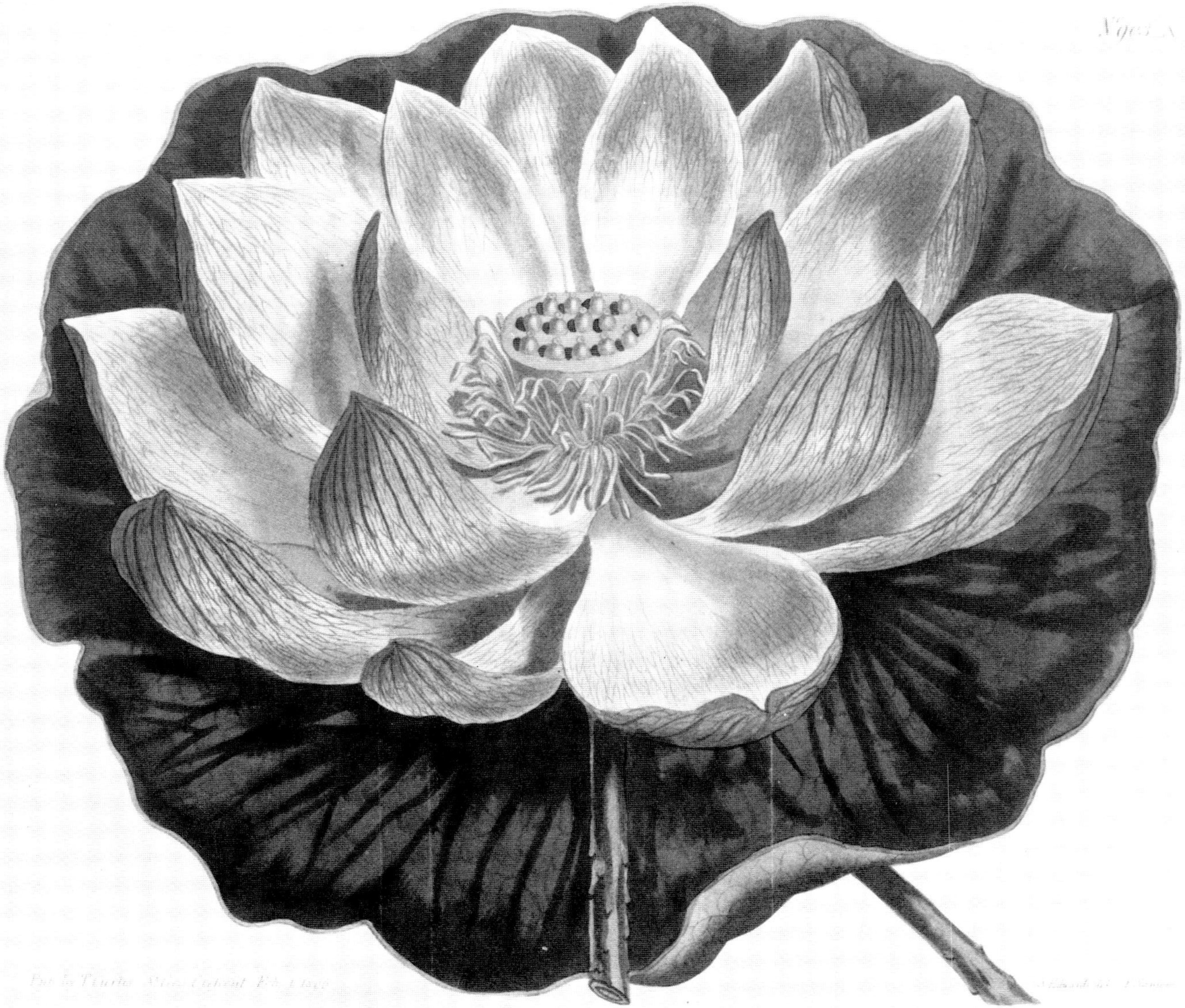

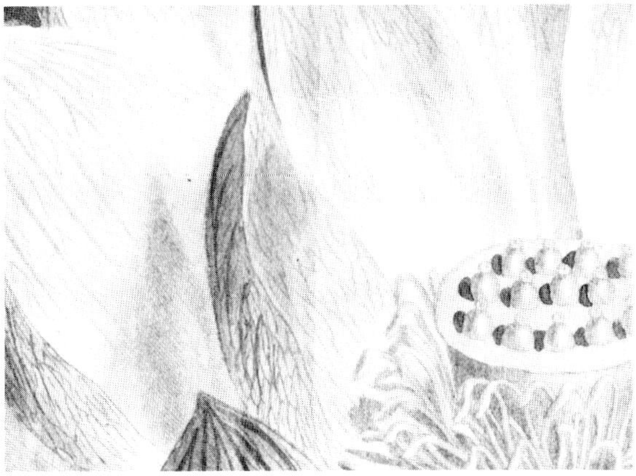

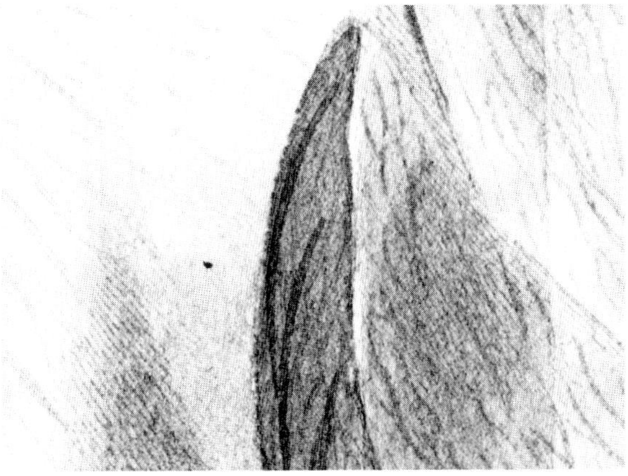

32. Nelumbium speciosum. Sacred Bean of India. (title on facing text page). Line etching, engraved by Francis Sansom (English fl. ca.1785-180?) after a drawing by Sydenham Teast Edwards (Welsh 1768-1819). Forming a full-page illustration printed at London in 1806 by Stephen Couchman. Printed in sanguine ink (platemark 20 x 23.2 cm.) on wove paper (wire side, 22.8 x 28.2 cm.). Hand-finished with transparent watercolor. Ex: J. Sims. *Curtis's Botanical Magazine:...continued by John Sims.* London. Volume 23, 1806. Plate 903A. Nissen BBI #2350.

The majority of hand-colored plates were finished with thin transparent watercolors painted over black-printed images as in the Flora Londinensis *example. The special luminosity of watercolor is created by light being reflected back through it from the paper beneath. Prints prepared with the intention of final hand-coloring therefore contained plenty of white space; outline was precise and often quite detailed, but shading was suggestive rather than elaborate. The colorist simply painted within the well defined outlines of most of the vegetation with broad washes, adding touches of deeper or contrasting colors for shading. The flowers were more demanding, but if the printmaker's line adequately depicted structure and perspective, the colorist's task was kept fairly straightforward. The work called for good eyesight, deftness and time-consuming patience.*

Occasionally, the printed black line clashed with pale coloring or simply created an inaccurate impression. This illustration exhibits one solution to that problem. It was prepared for Curtis's other great publication, the Botanical Magazine, *in which the attractive coloring of exotics was important to commercial success. In order to heighten the roseate coloring of the large, plump showy petals, the plate was printed in a pale russet ink. The coloring was obviously perfunctory, but the harmony of ink and coloring is immediately apparent and quite pleasing. If printed in black, the etched network of venation on each petal would have given the bloom a rather murky cast.*

This illustration was prepared for an inexpensive, mass-produced magazine that published, on average, 45 plates a year in a large edition (the first part, published at one shilling in 1787, sold 3000 copies). The use of colored ink in this way was quite common in this and other botanical publications for which the cost of hand coloring had to be kept within tight bounds and life colors were therefore only suggested rather than elaborated.

Many illustrated flower books were republished in later, cheaper issues. Since reissue success was based on past reputation, the publisher sometimes paid closer attention to profits than to niceties of production. It is not uncommon to find that a reissue of this type of print was carelessly done in black and that the quality of the coloring was also much inferior to that of the first version.

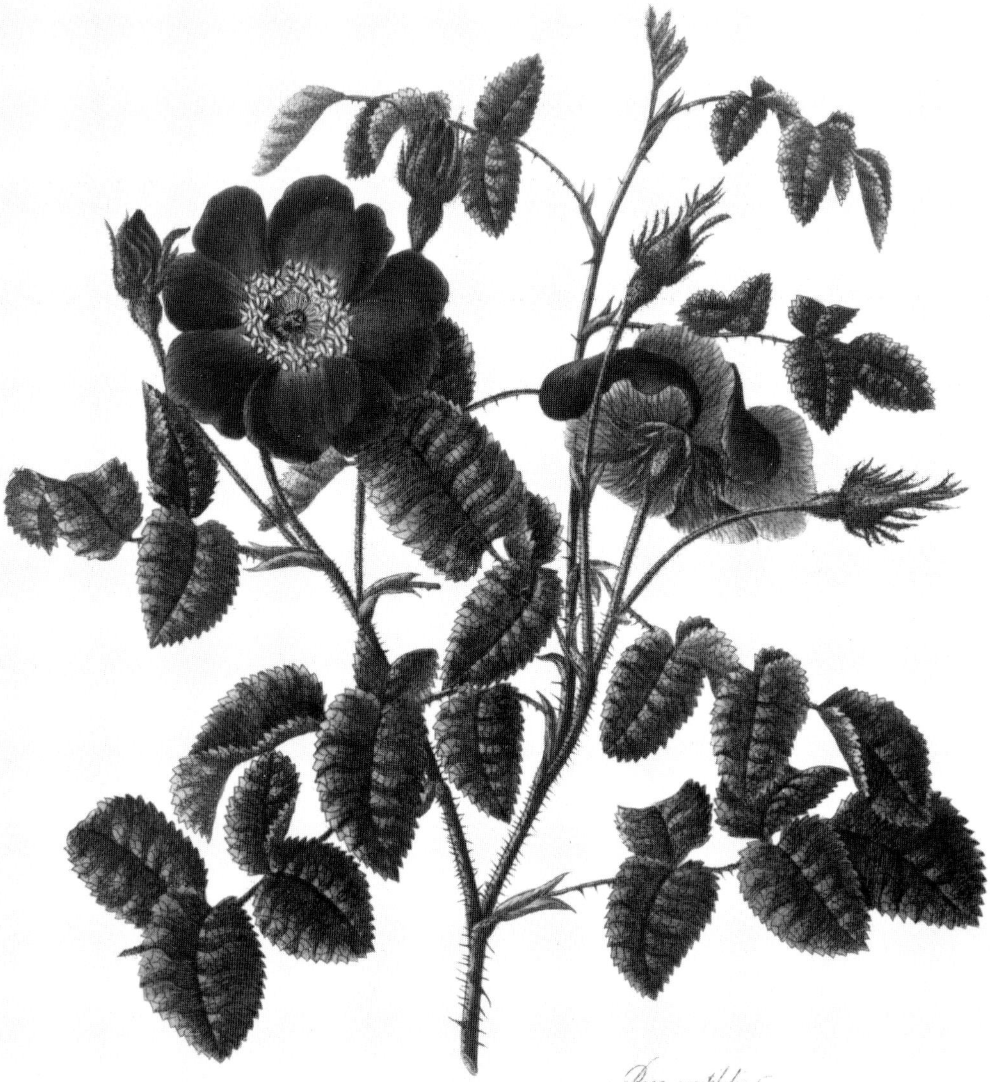

Rosa centifolia E.
Single Velvet Rose.

Published by Mary Lawrance Teacher of Botanical Drawing &c. [illegible] Queen Ann Str. East London [illegible]

[illegible] by Mary Lawrance

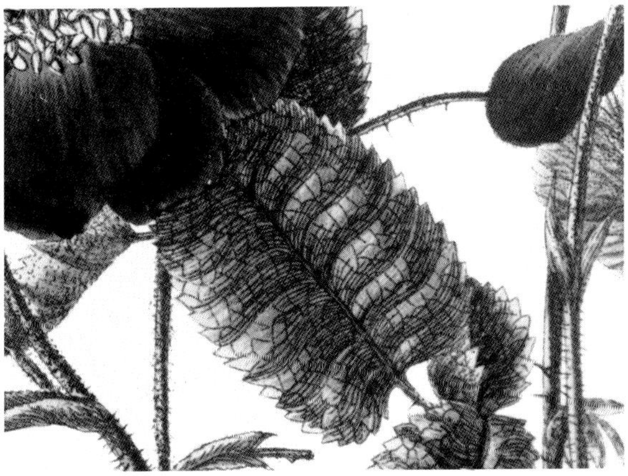

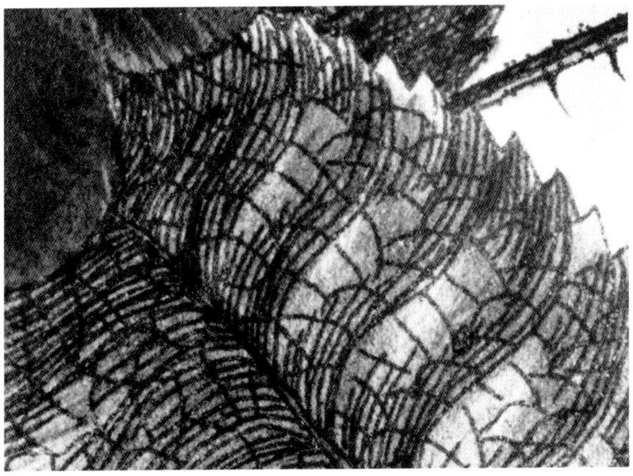

33. Rosa centifolia. E. / Single Velvet Rose. Line etching, with parallel-line shading, some local cross-hatched shading, dotted line and stipple, drawn and etched by Mary Lawrence (English 177?-1830). Forming a full-page illustration printed at London in 1798, plate printer unknown. Printed in black ink (plate-mark 33.3 x 27.8 cm.) on wove paper (wire side, sheet size 47.2 x 32.5 cm.). Hand-finished with heavily applied transparent and opaque watercolor, heightened with gum or albumen. Ex: M. Lawrence. *A Collection of Roses from Nature*. London, 1799. Plate 59. Nissen BBI #1151.

The language of the printed image in a transparently colored illustration is essentially the major source of information. It is still clearly "readable," but if coloring is opaque, the printed lines and tones become muddled and lose a lot of their meaning. However, transparent coloring naturally tends to be pallid, not only in intensity but also in emotive appeal. Some artists preferred dense or quite opaque coloring and were prepared to sacrifice the language of the print to the vividness of their painting, treating its lines and tones as little more than a guide to the form of their superimposed picture. In such a situation, the colorist's work obviously assumed a dominant rôle in the scheme of the illustration. Prints deliberately prepared toward this end often exhibit great economy in line and shading.

Miss Lawrence's ambitious work on roses boasts no fewer than 90 plates, all finished with heavy, intense coloring. Each print is signed as having been drawn and etched by herself, a "Teacher of Botanical Drawing." However, it is the coloring that gives them their appeal, and by a happy accident its intensity has obliterated most of the etched line beneath. Teacher of botanical drawing she may have been, but her etchings were apparently unhindered by any understanding of the botanical structures she depicted. Her technique, Dunthorne laments, "...is very poor, especially of stipple used for some blooms, which entirely fails to express any modeling or structure of the flower"—no better, he finds, than "the meaningless network of lines with which the leaves are composed." These brightly colored pictures were clearly intended to please lovers of roses, who obviously included the dedicatee, Queen Charlotte, rather than to add to botanical knowledge.

Miss Lawrence was her own colorist and applied her colors with painstaking care. Apparently undaunted by the Herculean task of doing the whole edition of 90 plates, she never allowed her heavy painting to stray across a line, and her colors, all prepared under her supervision at her own house, were consistently vivid and eye-catching. Nevertheless, the overall result has a certain crudity, and her greens, though bright, are distinctly monotonous. This example is specially selected from the group because it reveals some of her idiosyncratic printmaking technique that is usually obliterated by the coloring.

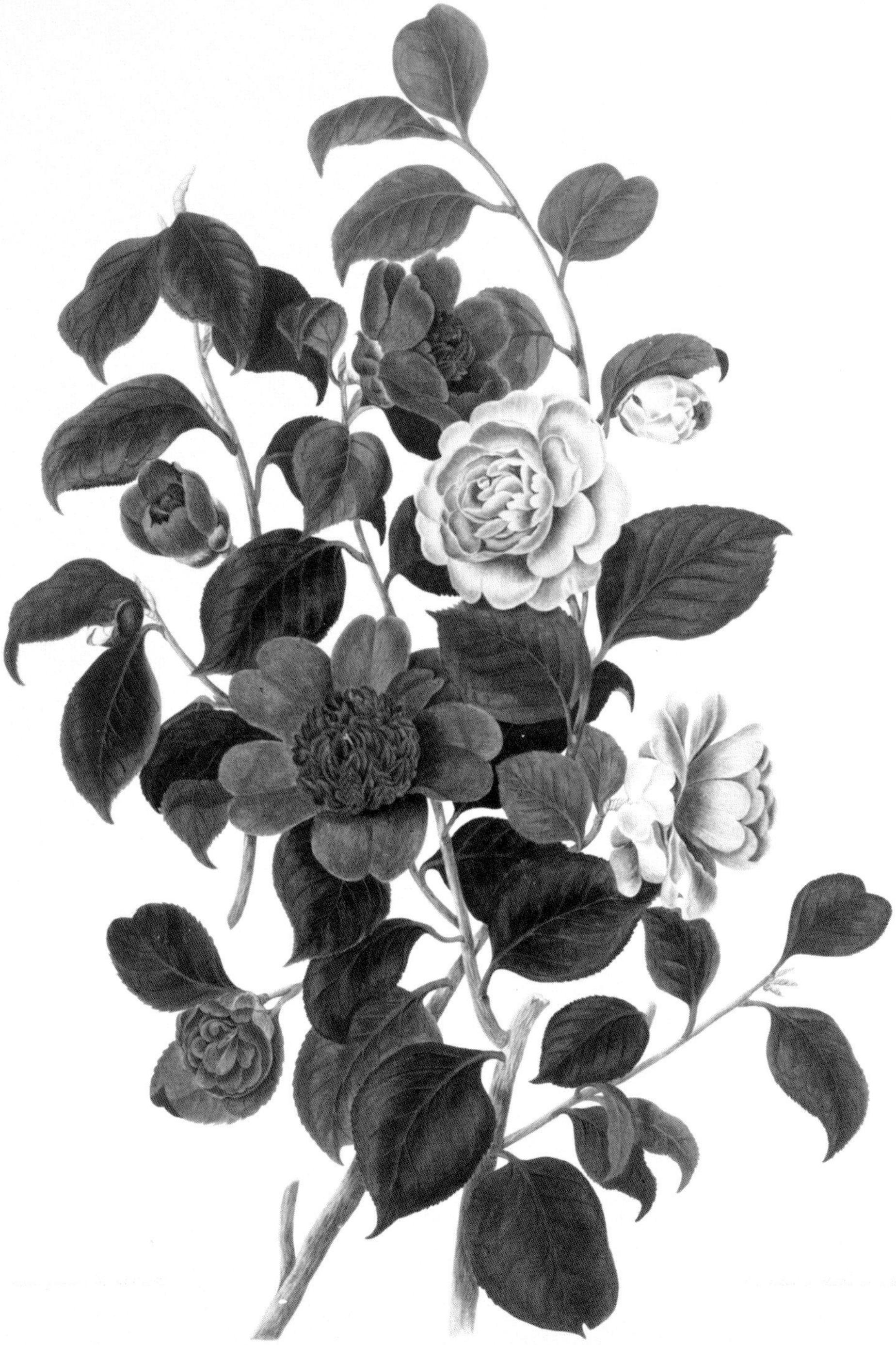

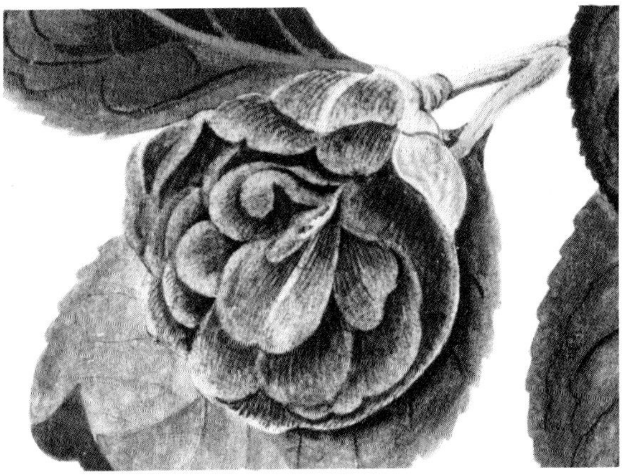

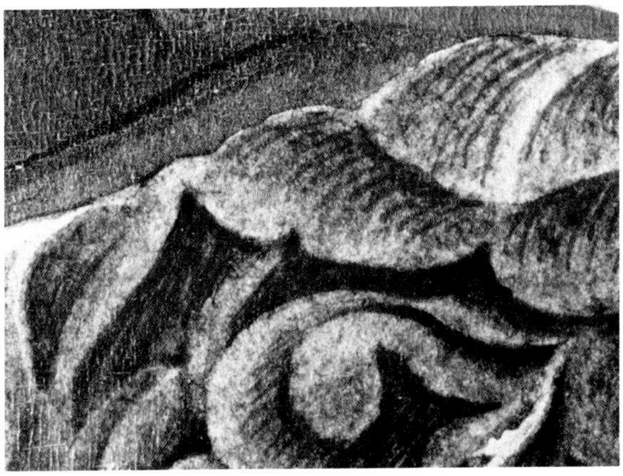

34. Anemone flower'd or Waratah Camellia. Rose Color'd or Middlemists Camellia. Line etching with pale aquatint, engraved by Weddell (possibly H. H. Weddell, English fl. 1819-1840) after a drawing by Clara Maria Pope (English 1750?-1838). Forming a full-page illustration printed at London in 1819, plate printer unknown. Printed in black ink (platemark 58 x 47 cm.) on wove paper (wire side, watermark "J. Whatman 1818," sheet size 62.3 x 49.5 cm.). Hand-finished with heavy opaque watercolor, heightened with gum or albumen. Ex: S. Curtis. *A Monograph on the Genus Camellia.* London, Arch, 1819. Plate [2]. Nissen BBI #437.

Twenty years after Miss Lawrence's Roses *was published, another lady artist launched some even more ambitiously colored prints onto the market. Clara Pope was, however, an artist of much better calibre. Her large and very impressive studies of camellias were prepared for a monograph on the genus by Samuel Curtis, much younger cousin of William. Sadly, the work was abandoned after only five prints had been published and the drawing made for a sixth. These imposingly large prints were etched and delicately shaded in aquatint by one of the Weddell family of engravers.*

Blunt comments that Mrs. Pope "had a sense of the dramatic and knew how to paint in the grand manner," as this example bears out. She painted a flower portrait over Weddell's print using mostly heavy and opaque colors. In fact it is quite difficult to discern what kind of print lies beneath, leaving us to wonder whether his delicate work was necessary if her intention was to forcefully obliterate it! The glossiness of camellia leaves is not easily captured with opaque watercolor so the artist gave them a heavy gum arabic overglaze.

The local use of gum arabic or albumen can be found on a great many hand-colored botanical prints. This gives a certain lustre and depth to an otherwise cloudy or "dry-looking" color. An alternative practice was to burnish some darker colors with the brush handle. Like the practice of using a colored ground ink, these were inexpensive means of enhancing the look of initially costly hand coloring.

Although an attractive selling feature, hand coloring could greatly increase the cost of illustrated book production. For example, colored copies of Curtis's Flora Londinensis *cost twice the price of plain, and specially colored copies three times as much. That means that the combined expense of paper, engraving, type-setting, printing, folding and sewing was matched, or even exceeded, by the cost of this final embellishment. The early demise of this lavish camellia monograph may well have been occasioned by the high cost of its coloring.*

Pinus Sylvestris Pinaster vulgo.

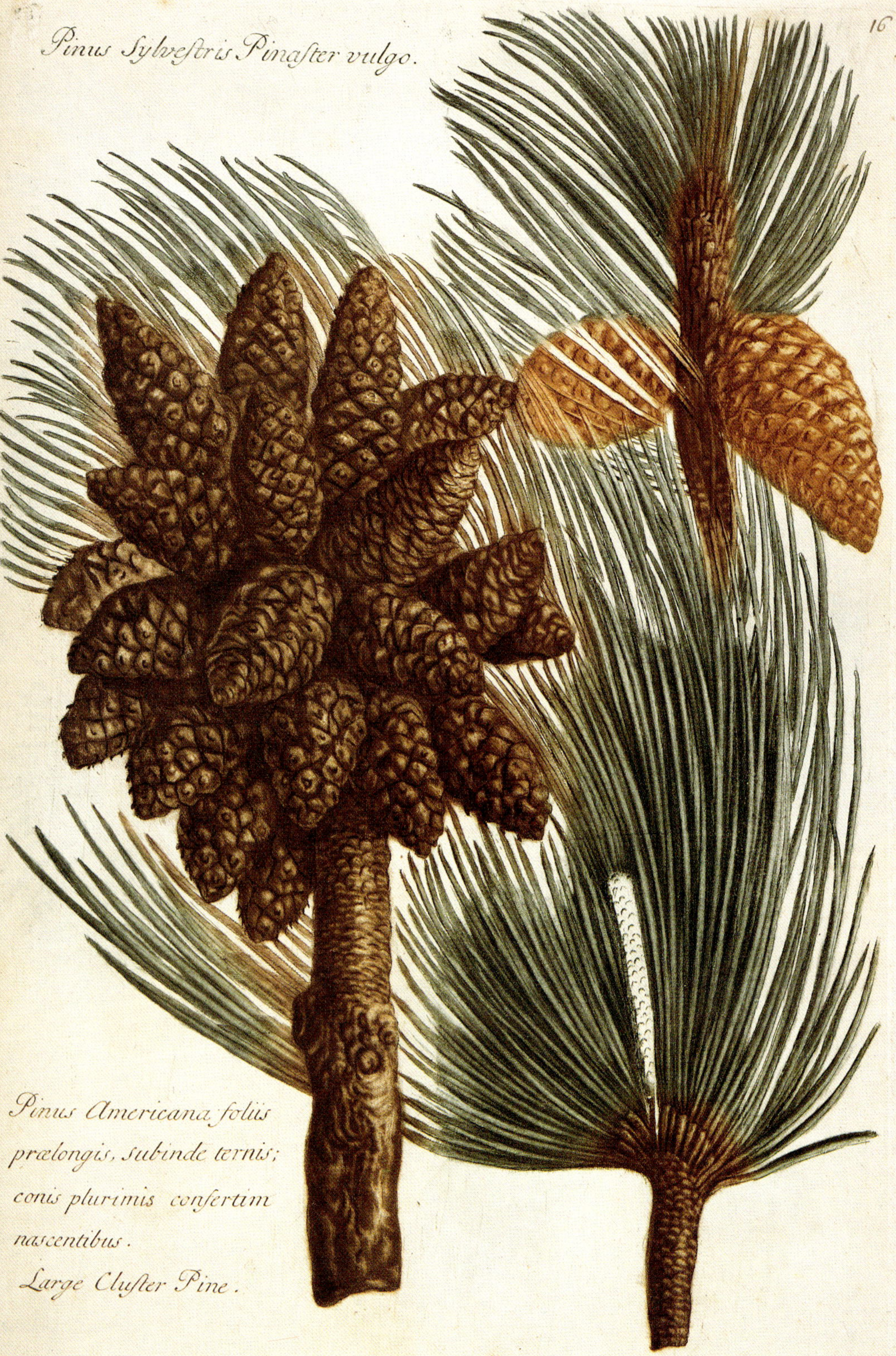

Pinus Americana foliis prælongis, subinde ternis; conis plurimis consertim nascentibus.

Large Cluster Pine.

Van Huysum pinx.

E. Kirkall sculp.

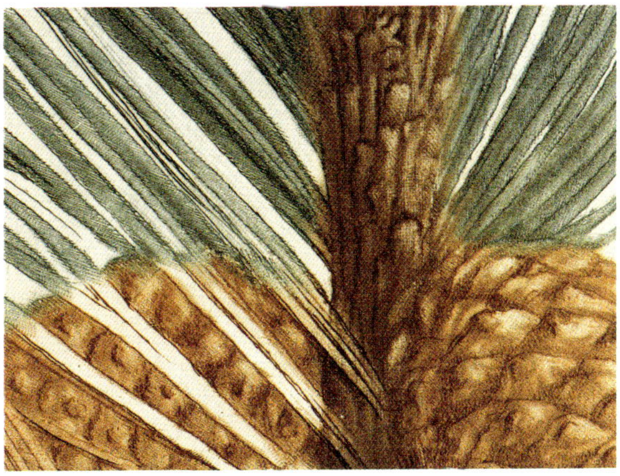

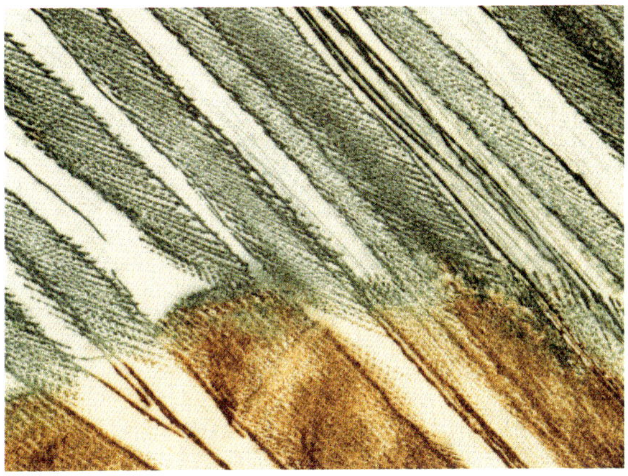

35. Pinus Sylvestris Pinaster vulgo. Line etching, with rocked tints and locally scraped highlights, engraved by Elisha Kirkall (English 1682?-1742) after a drawing by Jacob van Huysum (Dutch ca.1687-1746). Forming a full-page illustration printed at London in 1730 by Kirkall. Color-printed, à la poupée, in green, umber and sienna inks (platemark 38 x 25.5 cm.), on laid paper (wire side, watermark a "T" with small circle on stem, page size 44.7 x 28.8 cm.). In: Society of Gardeners. *Catalogus Plantarum...A Catalogue of Trees, Shrubs, Plants and Flowers, ...which are Propagated for Sale by a Society of Gardeners.* London, [1729-]1730. Plate 16. Nissen BBI #2230.

A solution to the unacceptable degree of individual variation in hand-colored illustrations was to print the colors. Printer's inks, usually mixed in sufficient quantity to feed a long run or a whole edition, generally retained uniformity of color from first to last. The chief shortcoming was the limited choice of pigments before the introduction of aniline dyes in the mid-19th century. In fact, although color-printed botanical prints were produced in growing numbers from the early 18th century onwards, they never wholly displaced hand-colored ones until the end of the Victorian era.

Our earliest example of color printing, dating from 1730, is the work of the Englishman E. Kirkall, a pioneer experimenter with intaglio color printing who even tried combinations of intaglio and relief. He had particular success in creating tinted pictures with a combination of etching and mezzotint. He line etched the subject in some detail, used a mezzotint rocker to superimpose tones over the required parts, and finally burnished out the highlights and any tint that had strayed over an outline. He inked selected parts of the image with appropriate colors, and the remainder with the dominant color, usually dark green, before printing the whole at one impression. At this early period the range of ink pigments was apparently very limited, and Kirkall rarely used more than three colors at a time.

Kirkall prepared color plates for two major botanical publications, one of which is shown here. The other work, Martyn's Historia Plantarum Rariorum *(1728 [-1732]), included 50 plates by Kirkall's process; this one has only seven. His novel method, undoubtedly slow to prepare and probably costly to print, must have held great interest for the connoisseurs of his day. To our eyes the same monotonous green ink seems to pervade all, and the second and third color details are often blurred and rarely as convincing as hand coloring. In fact, a number of his prints have final details added by hand. Most of the plates in this* Catalogus *are hand-colored engravings; only a third are Kirkall's work. Perhaps the slowness or cost of his process exceeded the patience or purses of the nurserymen who underwrote the preparation of this "trade" publication. In any case, Kirkall's work forms an early landmark in the history of color printing, which was to find some of its greatest challenges and successes in the reproduction of botanical illustrations.*

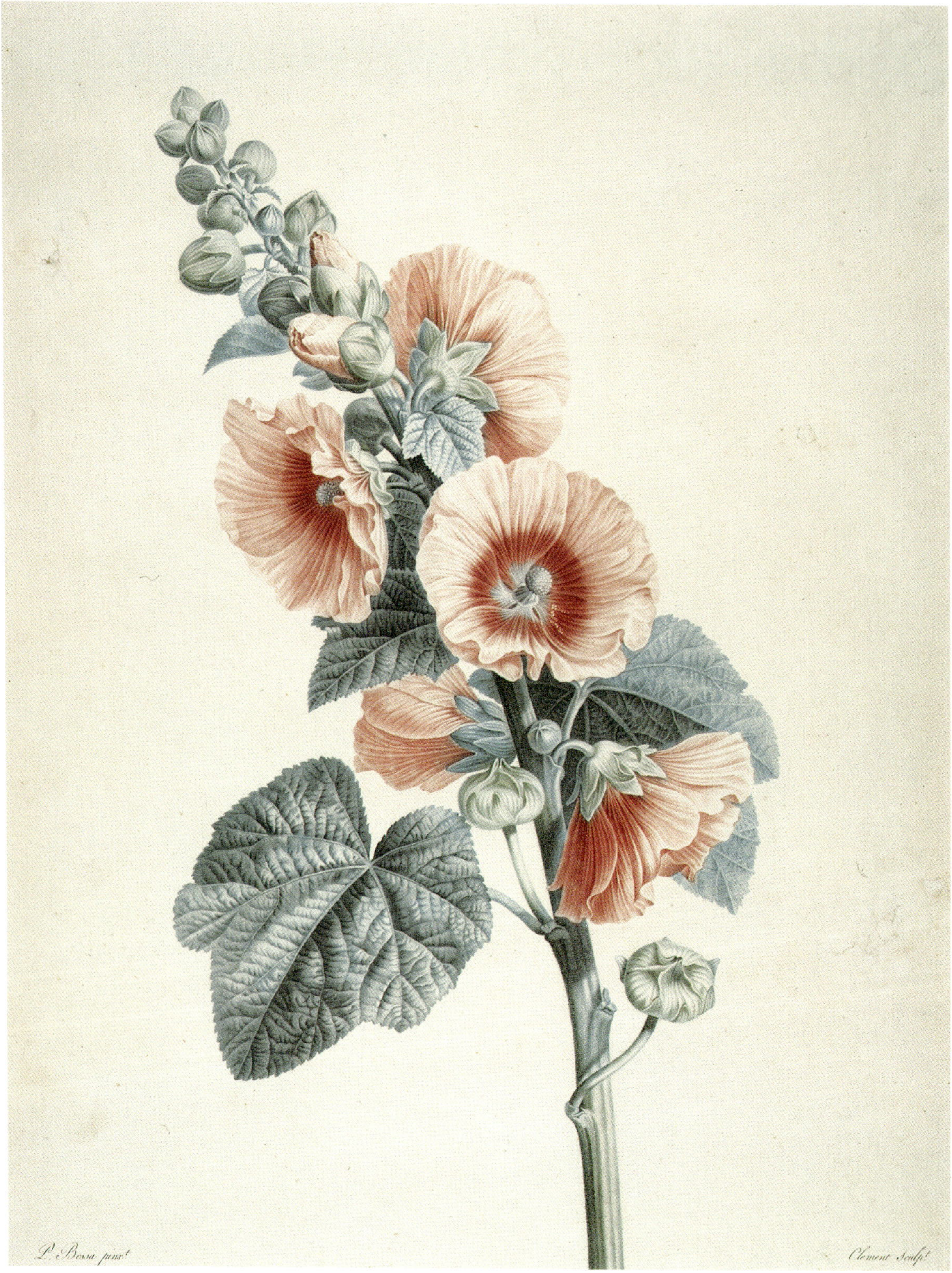
P. Bessa pinx.t
Clement Sculp.t

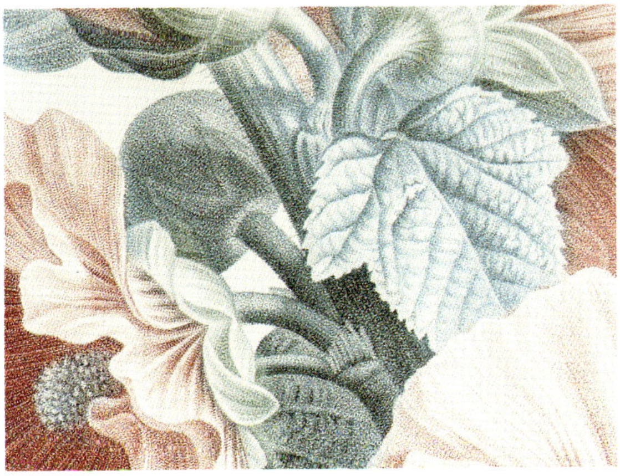

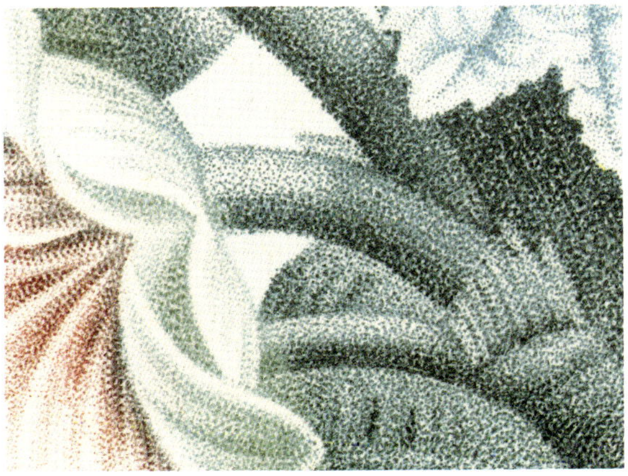

36. [Althaea rosea]. Stipple etching, engraved by Alex. Clement (French 1775-1808) after a painting by Pancrace Bessa (French 1772-ca.1835). Forming a full-page illustration printed at Paris in 1808, plate printer unknown. Color-printed, à la poupée, in dark and light green and red inks (platemark 50.5 x 37.5 cm.) on wove paper (sheet size 51.5 x 38.5 cm.). Ex: P. Bessa. *Fleurs et Fruits*. Paris, 1808. Plate [?]. Nissen BBI # 160.

It was Kirkall's method of printing two or three colors from a single intaglio plate, rather than his combination of line and rocked tints, that presaged later successes in botanical color printing. Kirkall's use of locally applied colored ink was much elaborated by another Englishman, Robert Laurie, who described his process to the Society of Arts in 1776. He used either a stump-brush or a small knob of rag—à la poupée (i.e. shaped like a wooden doll's head)—to paint various colored inks onto details of the intaglio image. Laurie favored plates prepared in a combination of mezzotint, stipple and line, and with very careful inking and skillful wiping could print a multi-colored picture at one impression. His method never allowed the colored inks to overlap and intermix. This form of color printing, especially applied to stipple engraving, soon created a fashionable craze in London for color prints, and it was during this boom, in 1786, that Redouté visited there and learnt the process. In England, color printed stipple engraving was applied to the production of portraits, fancy prints and reproductions of paintings, but rarely to botanical work. Rather, Redouté's efforts back in France are what established its use for plant illustration with such conspicuous success, and indeed he claimed much personal credit for developing the process.

This example, engraved after a painting by one of Redouté's pupils, exhibits color-printed stipple engraving in its purest form. There is no hand coloring whatsoever. All color was achieved by a combination of refined modulations of the stippled tone and skillful application of the dark and light green and red inks. The printer clean-wiped the plate, thereby confining ink strictly to the engraved dots and left no surface tone. This allowed the whiteness of the paper to shine between the printed dots, and where the stippling is more delicate or scattered it lightened the tint.

This print makes a fascinating comparison with item #24, which depicts another species of mallow in a black-printed stipple engraving.

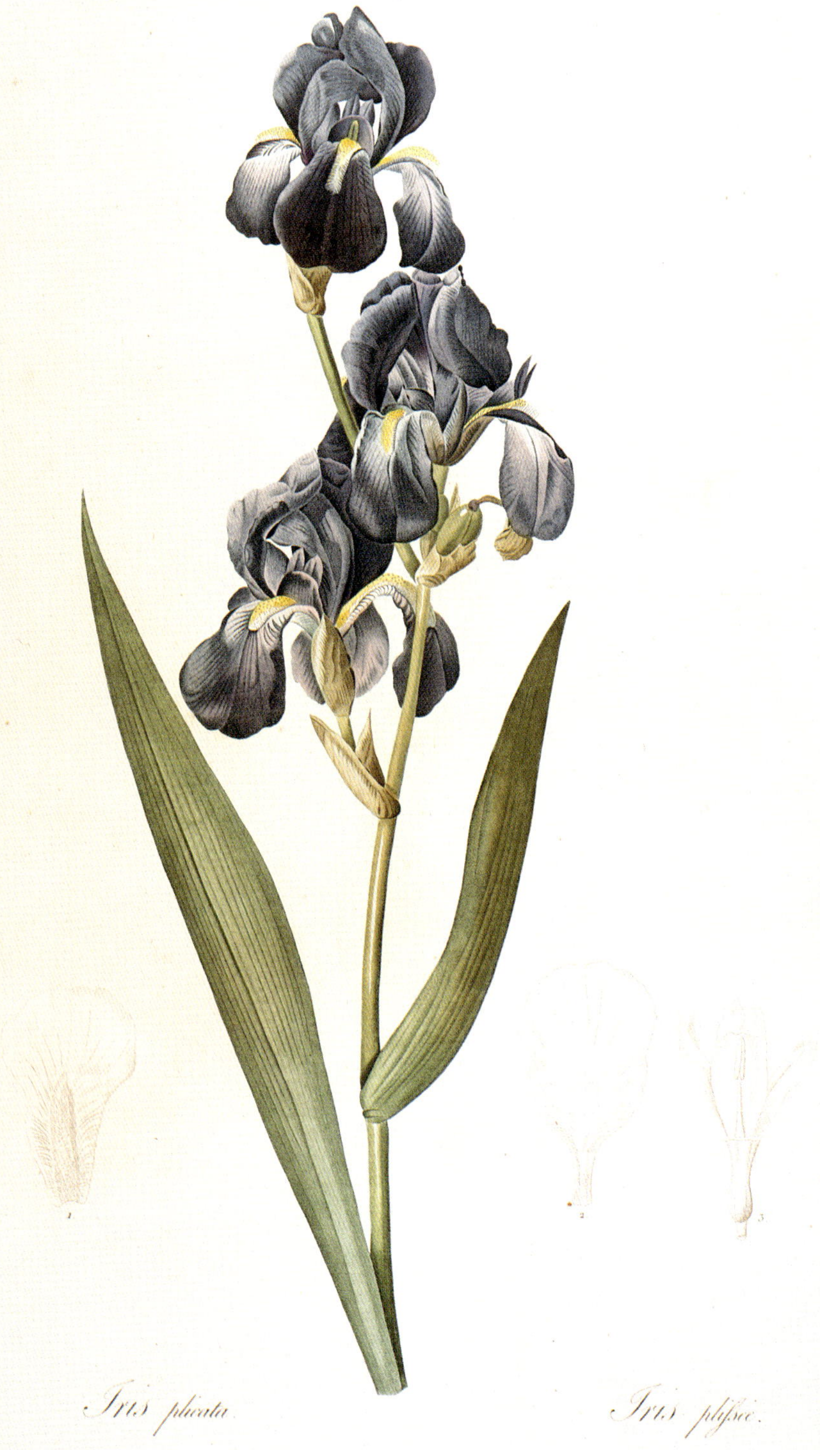

Iris plicata. *Iris plissée.*

P. J. Redouté pinx. Lemaire sculp.

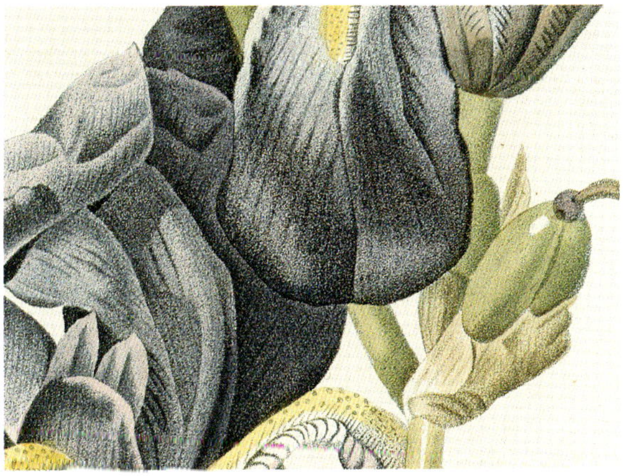

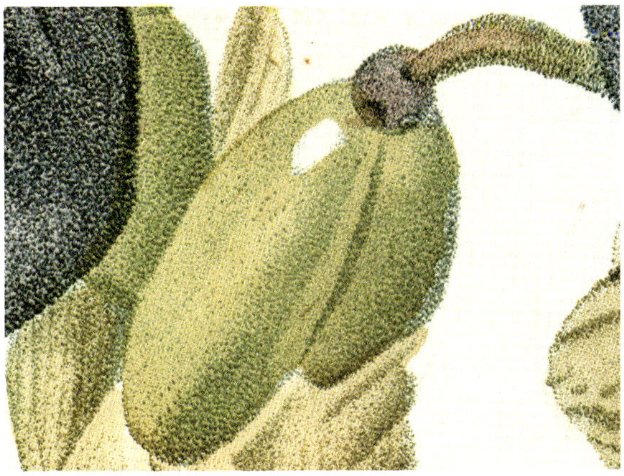

37. Iris plicata. Iris plissée. Stipple etching and roulette, with a little line engraving, engraved by [—] Lemaire (?French fl. ca.1800-1820) after a painting by Pierre-Joseph Redouté (Belgian 1759-1840). Forming a full-page illustration printed at Paris in 1812, plate printer unknown. Color-printed, à la poupée, in green, blue and umber inks (platemark 52 x 33.3 cm.) on wove paper (top side, sheet size 55 x 36 cm.). Ex: P. J. Redouté. *Les Liliacées.* Paris, chez l'Auteur, 1802-1816. Plate 356. Nissen BBI #1597.

The exceptional praise that Redouté's botanical art has earned for him is also due, in no small part, to the engravers who faithfully translated his watercolors into the prints by which the public knows him. Some 18 engravers worked on the 486 plates for Les Liliacées *over a period of 14 years.*

The subtle tones in Redouté's paintings were faithfully captured by his engravers. But the deftness of his linear brushwork in each tiny detail and the slight flamboyance that a generously loaded brush gave to his colors were somehow lost in the refinement of the engraving. The laying of dots with the burin and roulette was such a deliberately controlled process that it is scarcely surprising if the studied suavity of stipple fell short of capturing the spontaneity of brushwork. However, the results are magnificently skillful exercises in stipple engraving, highly refined in the treatment of shading and modelling. The balance of tones was exquisitely regulated so that, when reduced in value by color printing, they conveyed the correct sense of plasticity and perspective.

The finished effect almost invariably relied on a combination of stipple engraving and hand coloring. A thin wash of watercolor that closely matches the ink was applied over every part of the images with occasional use of contrasting highlight tints where necessary. The stipple engraver had to regulate his tones with this finish in mind, rather than rely on background whiteness of paper within the tint, which was used only to illuminate the bright highlights. The meticulous hand finishing must have added to the already high cost of producing the edition of 200 copies. An additional 18 copies were specially colored by Redouté himself.

This fine example is printed in green, blue and dark brown inks, and heightened with blending watercolors. It should be compared with item #25, a black-printed Redouté illustration, to appreciate the proportion of printed, as distinct from hand-painted, work.

Redouté had a special concern for color in his illustrations of this particular plant group: "...it is not only for the pleasure of the eyes that I have undertaken the...work; naturalists have long regretted their inability to conserve the Liliaceae in their herbaria; the accuracy of the descriptions and the fidelity of the engraving will save them the trouble of trying."

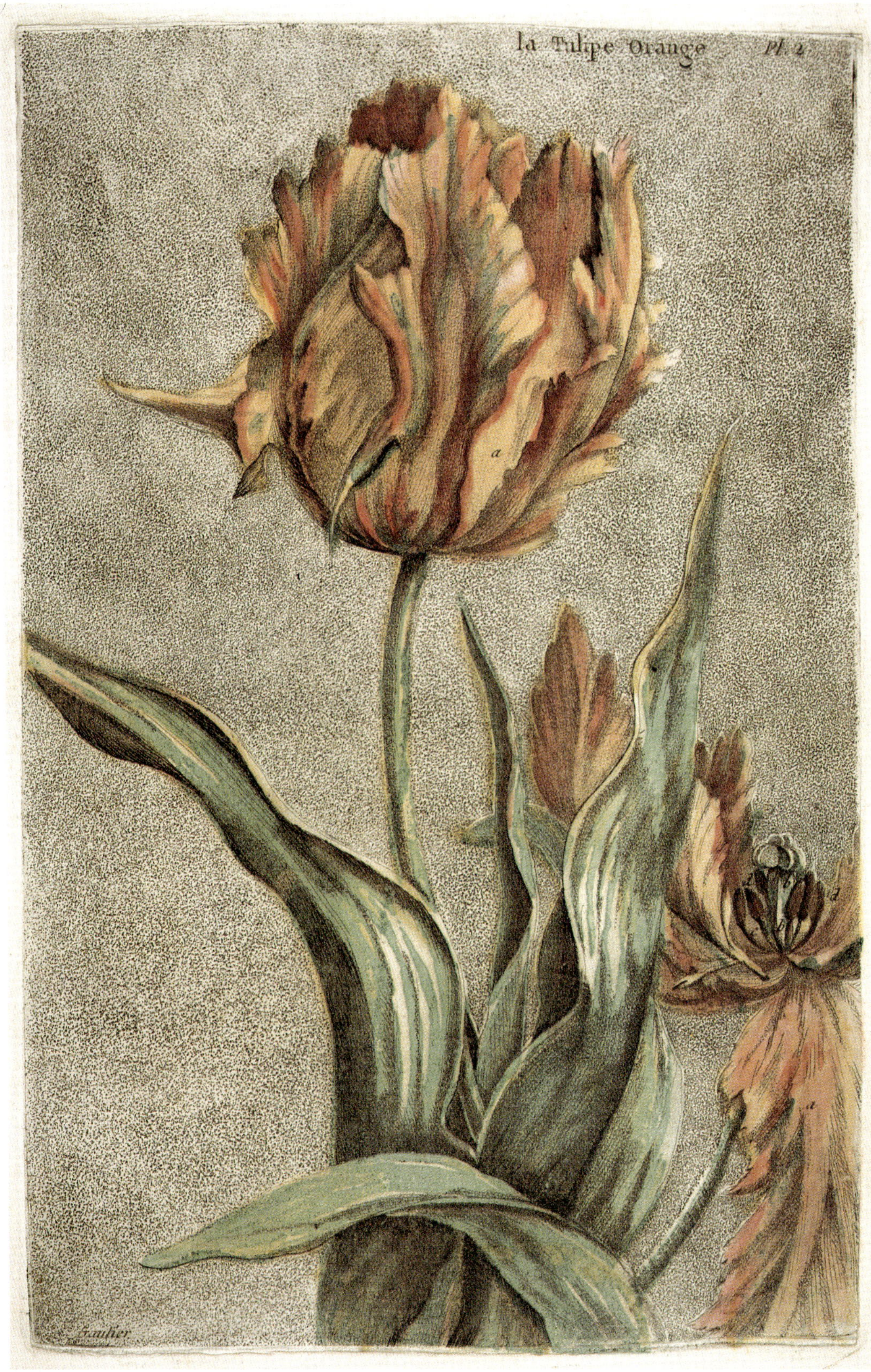
la Tulipe Orange
Pl. 2
a
d
a
Gautier

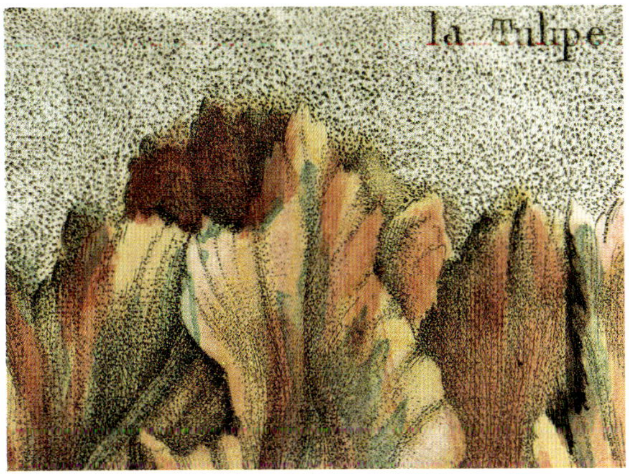

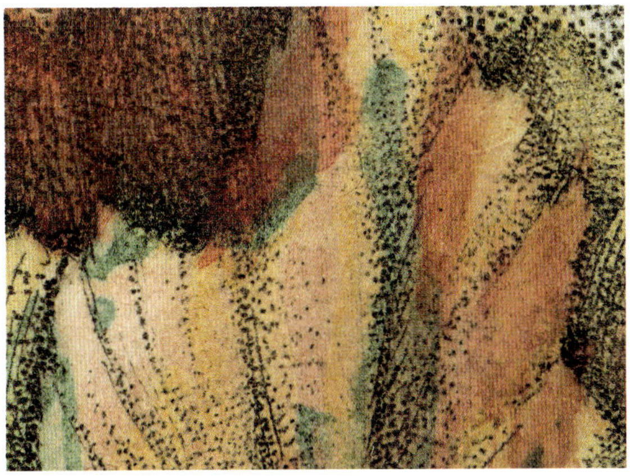

38. La Tulipe Orange. Line- and stipple-etched key plate, with rocked tints on three color-separated plates, engraved by Jacques Gautier-Dagoty (French 1717-1786). Forming a full-page illustration printed at Paris in 1767 by Gautier-Dagoty. Color-printed from a key and three separate tint plates (Le Blon's process) in black, green, ochre and red inks (platemark 29 x 19 cm.) on laid paper (top side, 41 x 28.5 cm.). In: J. Gautier-Dagoty. *Collection des Plantes Usuelles, Curieuses et Étrangères.* Paris, Auteur & Boudot, 1767. Plate 34. Nissen BBI #693.

À-la-poupée color printing relied on a choice of single-color inks carefully applied to an intaglio plate so that they did not overlap or intermix. Long before that, though, it had been recognized that all the hues of the natural world are merely combinations of primary colors. Newton had put this idea forward in the 17th century, suggesting that there were about seven primary colors. Others differed in their analysis of the primaries, and as long ago as 1704, J. C. Le Blon, a French printmaker working in Germany, produced colored mezzotints from a combination of only three color-separated plates, blue, yellow and red. He produced some astonishingly successful examples, and printed separations to demonstrate how the three-color principle worked. He was nearly 200 years ahead of his time, foreshadowing early 20th-century developments in trichromatic process-engraved color printing.

Le Blon was never successful in business, and after many vicissitudes he died in France in 1741, the rights to his invention passing to a former student, Jacques Gautier-Dagoty. In his later years the inventor had begun to experiment with the addition of a black "key" plate to enhance the body of his images. Gautier-Dagoty developed this aspect and claimed to have invented a new process thereby. However, he lacked Le Blon's finesse, and sadly mishandled the subtleties required for the process. He also apparently failed to understand the basis of Le Blon's three-color principle. Although he printed in three colors with a black "key," he rarely used colors as primaries for the creation of secondaries. In this example he used red and yellow inks but failed to make good color separations; even where the red and yellow overlap, the former dominates the combination, which thus fails to produce a good orange. In this he neglected a fundamental principle that was to become an inseparable part of later multicolor printing.

This is typical of Gautier-Dagoty's work. Register is poor, the rather crude colors are not true to life, and oil from the colored inks has struck through each print as a disfiguring brown stain. The heavy stippled black background, apparently pleasing to the engraver, is common to all plates in this book. Nevertheless, this somewhat artless work carried the seeds of later progress in botanical color printing.

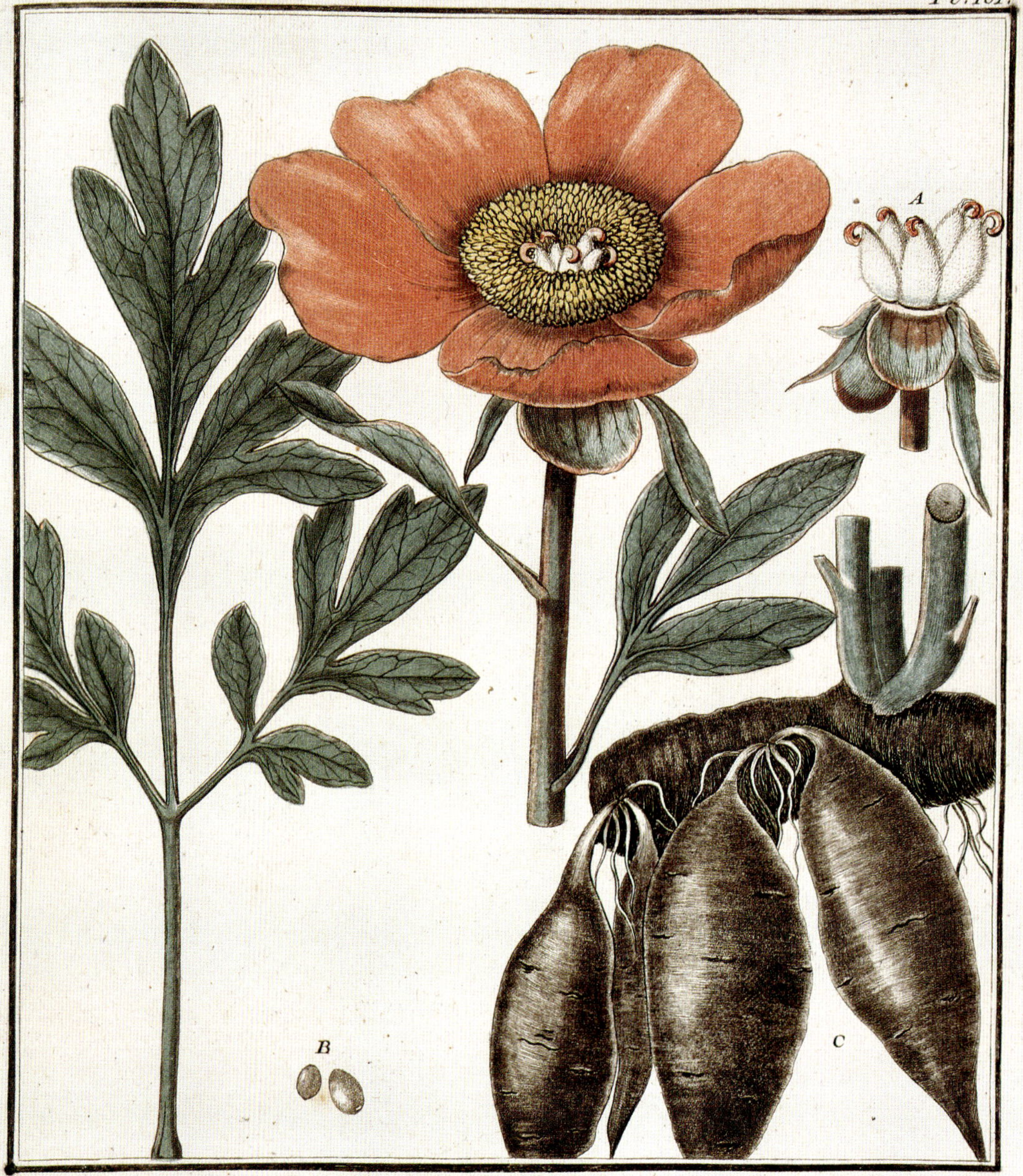

LA PIVOINE OFFICINALE FLO. FRANC.

Pæonia officinalis, *L. S.P. polya digy. 747. elle est vivace, elle croit spontanement sur les montagnes du Dauphiné et de la Provence, on la cultive dans tous les parterres, elle y fleurit en mai et juin.* **TIGES** *hautes de deux pieds ou environ, rameuses et rougeâtres à leurs extrémités supérieures. ses* **FLEURS** *sont grandes, solitaires, terminales, composées d'un calice à cinq divisions inégalles entre elles, de cinq pétales fort grands et d'un beau rouge plus ou moins foncé, d'un nombre indeterminé d'etamines et de deux à cinq ovaires chargés de poils blancs et terminés chacun par un stigmate aplati et coloré. ses Fleurs sont tres paßagères, il leur succède des* **FRUITS** *qui s'ouvrent et étalent un grand nombre de semences d'un beau noir, dans l'etat de maturité. ses* **FEUILLES** *sont pediculées et découpées profondémt. en lobes oblongs et élliptiques.*

N. B. Les jardiniers en distinguent deux variétés qu'ils nomment **MÂLE** *et* **FEMELLE**. *elle est ici representée un peu moins grande que de nature. A ovaires après la chute des pétales. B graines. C racines.*

sa racine quand elle est fraiche est âcre et amère, l'odeur narcotique de cette plante la rend suspecte; **M. DE HALLER** *la regarde comme vénenéuse, d'habiles medecins en ont proscrit l'usage et* **M. LIEUTAUD** *voudroit que par de nouvelles expériences on s'aßura mieux de ses vertus.*

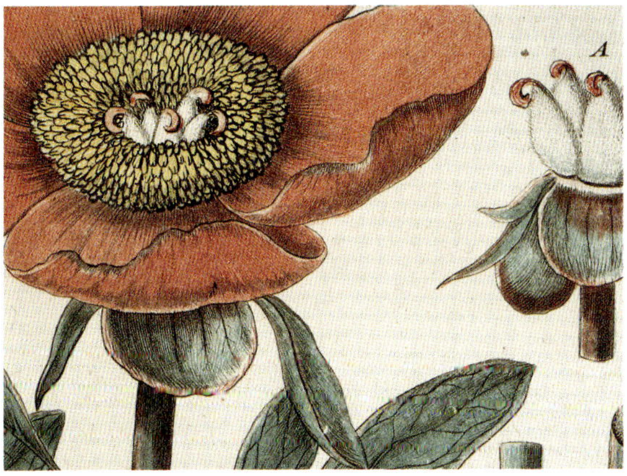

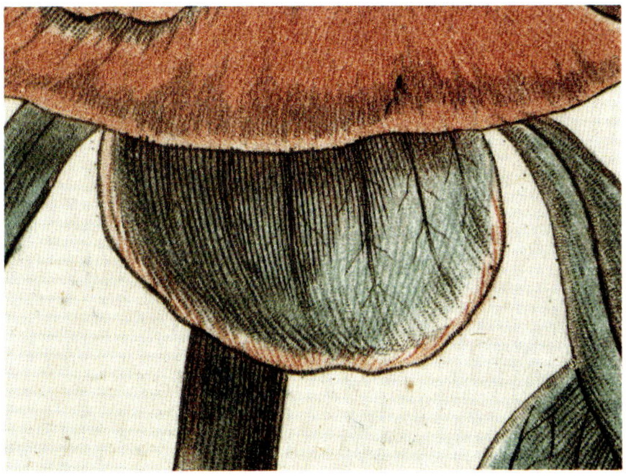

39. Plante Suspecte De La France. / LA PIVOINE OFFICINALE FLO. FRANC. Line-engraved key plate, with rouletted tints on three color-separated plates, engraved by Jean Baptiste François Bulliard (French 1752-1793). Forming a full-page illustration printed at Paris in 1787 by Bulliard. Color-printed from a key and three separate tint plates in black, green, red and yellow inks (platemark 22.5 x 16.5 cm.) on wove paper (top side, page size 35 x 25 cm.). In: P. Bulliard. *Herbier de la France.* Paris, Garnery [etc.], 1780[-1798]. Plate 101. Nissen BBI #296.

The potential of Le Blon's astonishing discovery of three-color printing was doomed to fade and disappear at the hands of Jacques Gautier-Dagoty and his five sons. Their inadequate craftsmanship was a major factor in this, but the family also held sole rights to the use of the four-color process and vehemently resisted any rivalry. In the event, other multiplate intaglio color-printing processes were developed and, in the hands of good craftsmen, were successfully applied to botanical illustration.

Pierre Bulliard's great Herbier de la France *is an admirable example. Its more than 600 plates, displaying rare craftsmanship, delicacy and accuracy, were all drawn, engraved and color-printed by the author. The whole book is a restrained masterpiece. The plates, which include a great many fungi, were engraved and printed with such accuracy that generations of French mycologists came to treat them as standard illustrations of their species. The flowering plants were similarly well-done, though the color of Bulliard's inks sometimes erred on the pale side.*

In this example Bulliard carefully line etched the outlines, venation and linear shading, and, in some places, used the roulette to create dotted, hence softer, lines. All this, together with the title, line frame and engraved text, formed the "key" plate, which was printed in black. He provided additional modelling by superimposing three tint plates, each engraved with the individual tones necessary to print separately the green, red and yellow that make up the image. He made the tints in each plate with the roulette, which he handled with masterly skill. The roots, being printed in black, most clearly demonstrate Bulliard's skill in the roulette manner of engraving. On certain other plates he appears to have created tints by extremely delicate use of the mezzotint rocker.

Bulliard was his own printer. He mixed colored inks with great delicacy and, in general, accuracy, and inked his tint plates with minute attention to detail. Some of his tinting seems to have relied on the additional effect of a rich ink tone to give greater local "solidity." The three tint plates were overprinted on the "key" plate with hair's-breadth accuracy, and evidence of his method of obtaining such precise positioning, or register, can be seen in pin holes at the upper right and lower left corners of the line frame.

The final effect is delicate, pleasing and botanically accurate, and the whole collection has a unique, quietly individual flavor. The degree of craftsmanship necessary to create these prints is only apparent when one studies them in close detail.

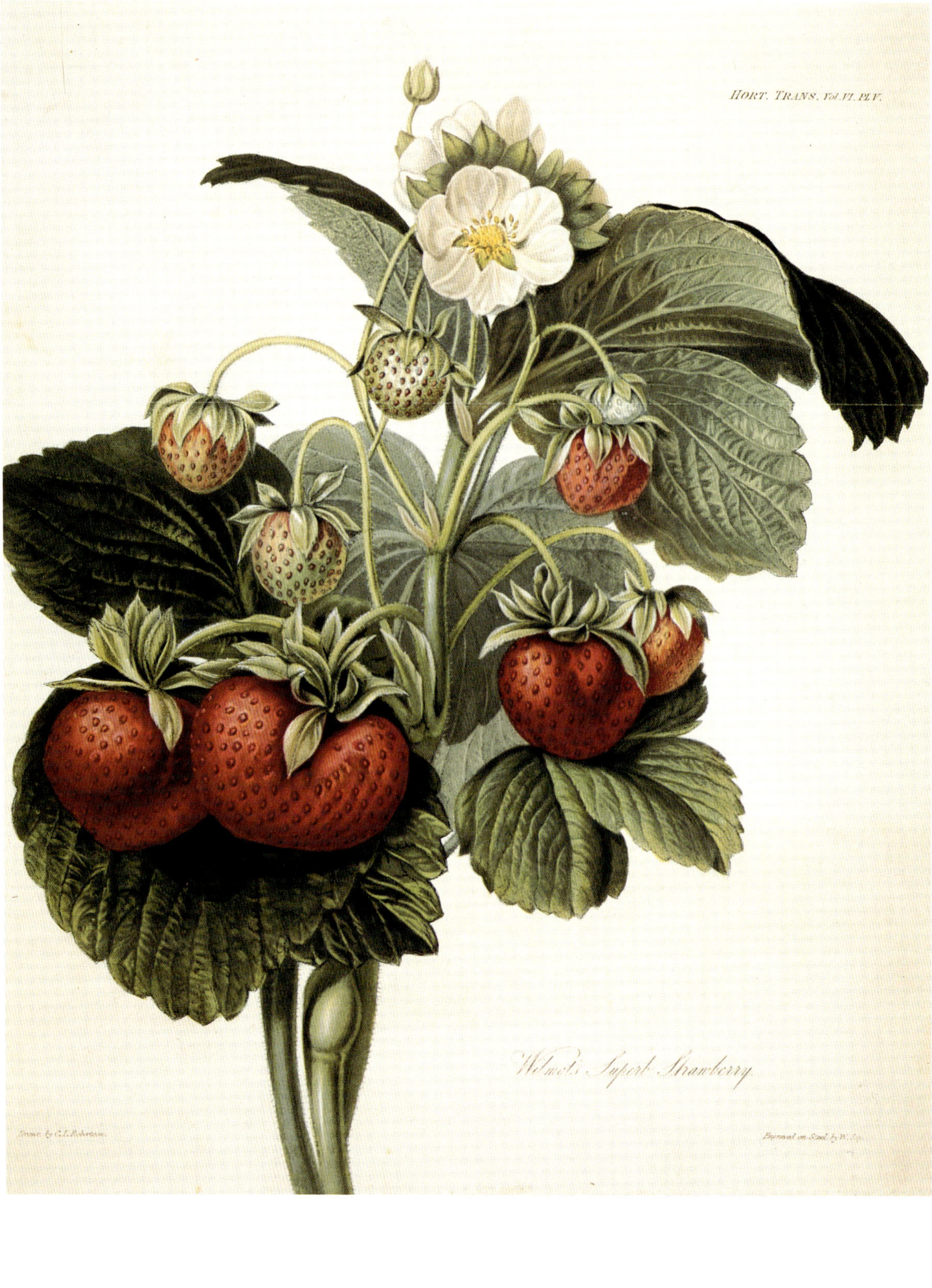

Wilmot's Superb Strawberry

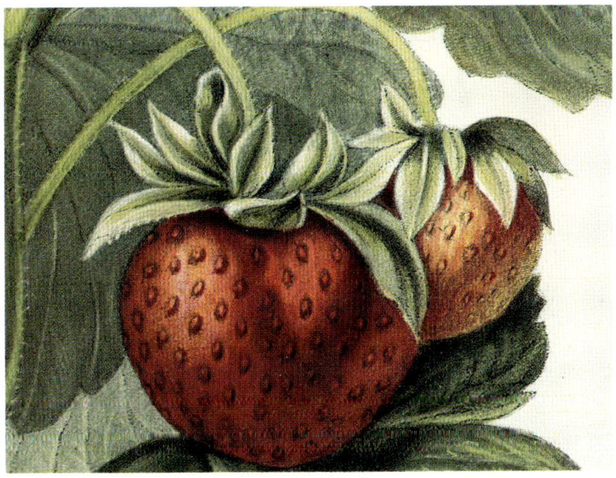

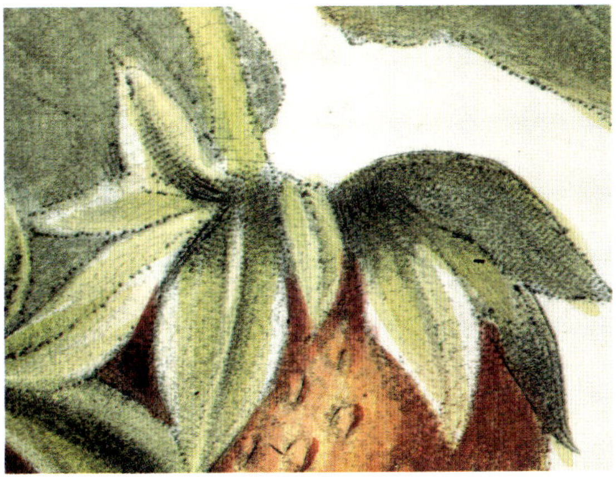

40. Wilmot's Superb Strawberry. Line-engraved and etched with rocked tints, "engraved on steel" by William Say (English 1768-1834) after a drawing by Charles John Robertson (English 1798-1830). Forming a double-page illustration printed at London in 1826, plate printer unknown. Color-printed, à la poupée, in blue and yellow inks (no platemark visible) on wove paper (sheet size 29.5 x 24 cm.). Hand-finished with transparent and opaque watercolor. In: Horticultural Society of London. *Transactions...* London. Volume 6, 1826. Plate 5. Nissen BBI #2387.

All previous intaglio exhibits, with the possible exception of item #12, were engraved or etched on copper, the standard material for intaglio printmaking for three and a half centuries. The number of prints that could be obtained from a copper plate engraving varied according to the hardness of the copper (the plates were hand beaten by the coppersmith) and the depth and manner of the engraving. Mezzotint plates produced the least number, possibly fewer than 70, while line engraving in the broad manner yielded the most, probably several hundreds.

In early 19th-century Britain there was a strong commercial interest in finding a material more durable than copper that could print much longer runs with little wear. After much experimentation, steel plates were produced that were soft enough to allow burin and other kinds of engraving. Once engraved, the plate was hardened and unheard-of runs printed from it. With rapidly expanding home and overseas markets for prints and illustrated books, this was a timely development for publishers. Apart from the familiar uses for town and country views, portraits, reproducing paintings and the like, steel engraving was also employed for scientific illustrations.

One of the very first English engravers who, in 1820, tried out the newly invented steel plates was the gifted mezzotinter William Say. His first essay in the medium was, typically, a mezzotint portrait from which an extraordinary 1200 impressions were taken. Say later used his newly acquired skill in the preparation of a number of steel-engraved flower and fruit portraits for the Horticultural Society of London.

The hardness of steel plate made for very fine engraving, as this print shows. Say prepared it with fine engraved and etched line, and added shading with delicate rocked tints. The plate was printed à la poupée in blue and yellow ink to blend with the highly finished hand coloring. As a result, the quality of the engraving was somewhat eclipsed, but close inspection reveals that it provides a considerable amount of modelling and structural detail to the finished portrait.

Say's series of horticultural plates were all signed "engraved on steel"—he was obviously eager to declare the nature of his craftsmanship. His sophisticated engraving and the unusually vivid coloring make a very striking plate, but what is really significant is that, despite its delicacy, this steel plate would print many hundreds of impressions with scarcely any loss of detail or sharpness.

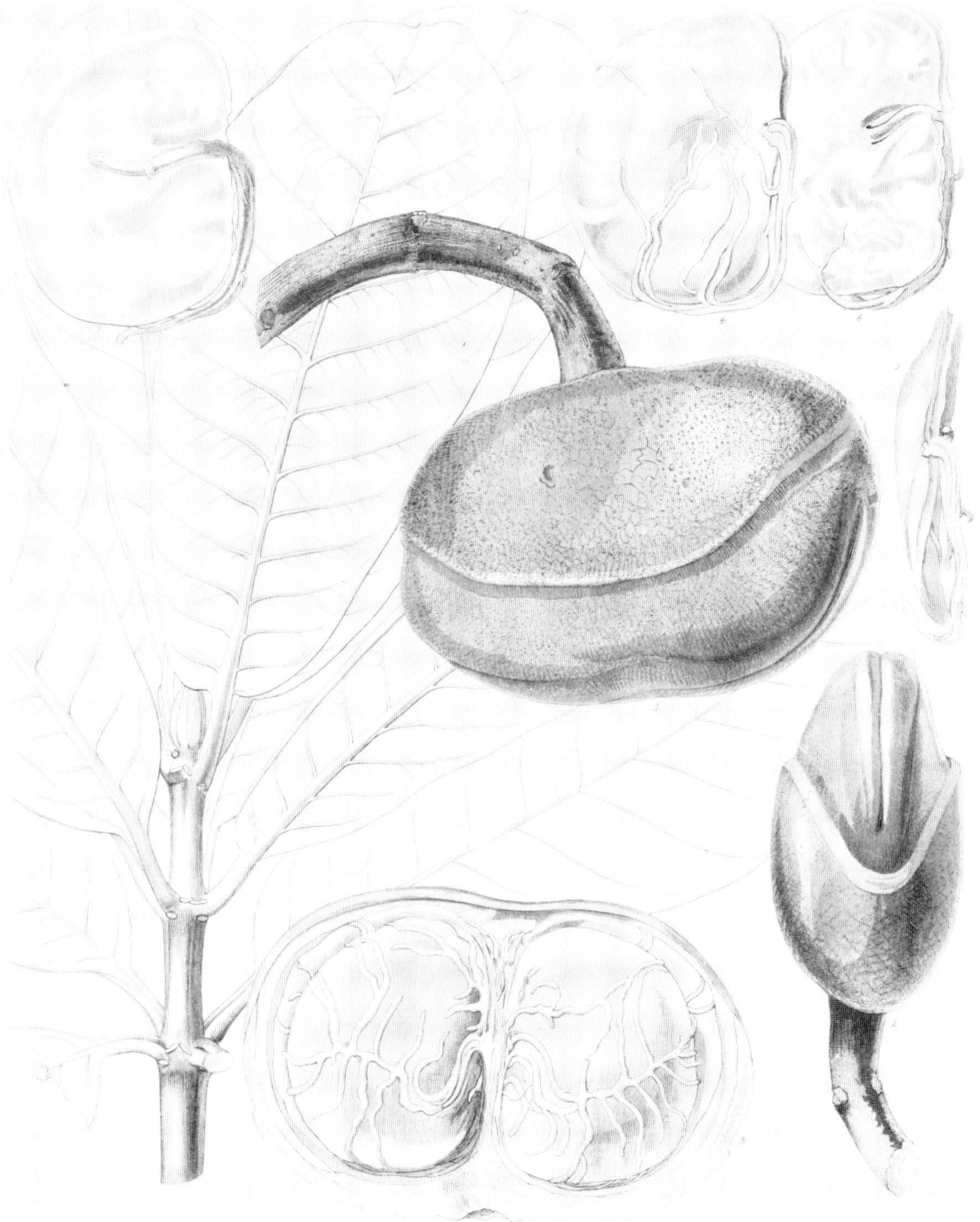

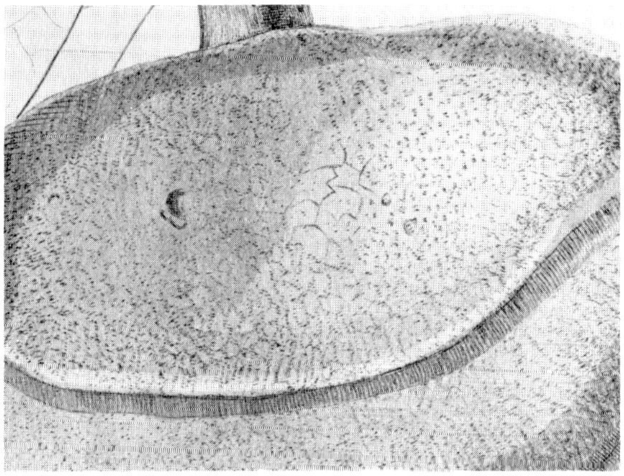

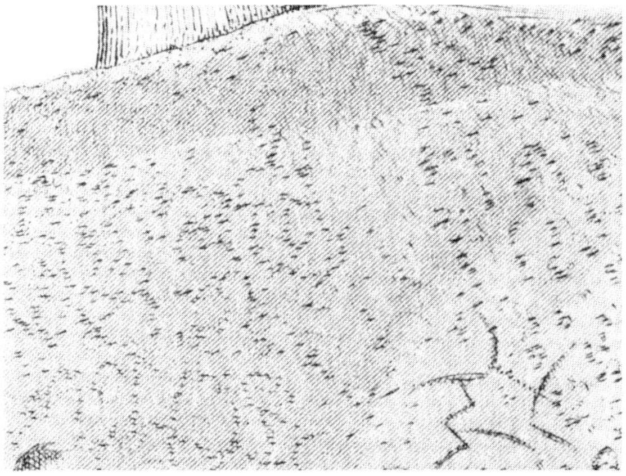

41. [Henriquezia obovata] (title from text). Line engraving, with cross-hatched shading and machine-ruled tints, engraved by George Jarman (English fl. ca. 1841-1880) after a drawing by Walter Hood Fitch (English 1817-1892). Printed at London in 1859, plate printer unknown. Printed in black ink (no platemark visible) on wove paper (wire side, page size 27.5 x 21.5 cm.). In: Linnean Society of London. *Transactions...* London. Volume 22, 1859. Plate 53.

Steel became widely accepted and used by engravers in Britain and elsewhere but, after the initial novelty, its use was rarely highlighted as Say had done. Apart from steel's great durability, the extremely fine work that could be engraved made it suitable for technical illustration where minutely detailed information was required.

The growing complexity of biology was mirrored in its illustrations, which became highly specialized, often depicting gross or fine structures that were meaningless to a non-scientist or, indeed, the average printmaker. Close cooperation between scientist and printmaker became essential if accuracy was to be assured, the former adopting an editorial rôle in the preparation of an illustration, the latter serving as specialized technician. George Jarman was a steel engraver who performed the latter rôle with distinction. Over a period of nearly 50 years, from the later 1830s onwards, he was an entirely unsung master of engraved biological illustration and did little work outside the science. About 180 full-page engravings in the Linnean Society of London's Transactions *are his work, one of which is exhibited here.*

This plate of the dissected Henriquezia *fruit is typical of the mid-career calibre of Jarman's work. Simply stated, it is a line engraving with cross-hatched shading and some ruled tints, but it deserves closer examination. By 1859, when this plate was done, chalk-style lithography had come into prominence for botanical illustration. Its autographic pencil-like character was specially associated with this style of botanical illustration, and particularly those lithographed by W. H. Fitch—who coincidentally made the drawing for this plate.*

Jarman was a line engraver and never deviated from that medium. However, he was eloquent in the language of his branch of engraving, and clearly chose to keep in step with the contemporary style. He therefore engraved a print that, at arm's length, has the look of a chalk-style lithograph but that repays examination with a powerful lens. Sturdy confident outlines were modelled with expressive cross-hatched shading and an incredibly delicate series of tints that were actually made with a ruling machine. He was thus able to include much more detailed information than the coarse chalk lithographic line would have permitted, whilst matching its broader tonal effects. The plate is a highly accurate and explicit scientific illustration, intentionally issued uncolored (the Linnean Society could rarely afford that cost) so that Jarman's fine monochrome tones would be allowed full expression. For this unpretentiously skillful illustration Jarman would have been paid about four guineas.

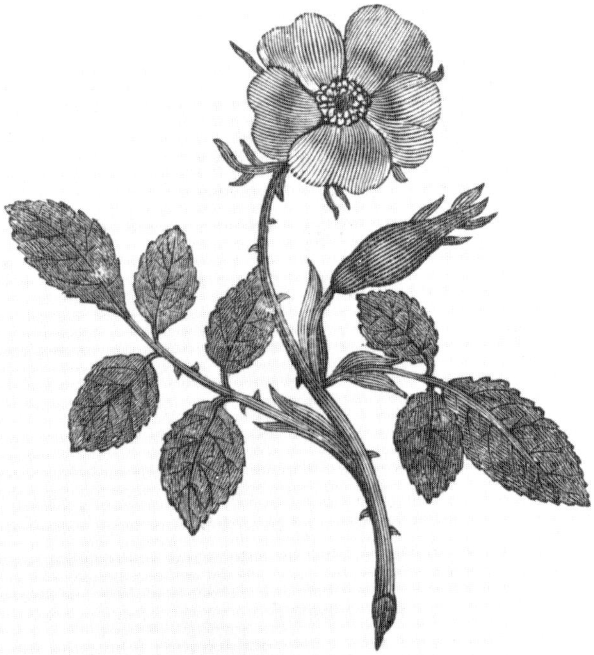

DOG ROSE.

ROSA CANINA.

Class XII. Icosandria. *Order* V. Polygynia.

ESSENT. GEN. CHAR. As the last.

SPEC. CHAR. *Germens* ovate: *Peduncles* smooth: *Stem* and *Petals* aculeate.

DESCRIPTION.

THE leaves are pinnate, composed of one, two, or three pairs of pinnæ, with an odd one at the end. The flowers are large and terminal, two or three together. The corolla is composed of five heart-shaped petals. The fruit is an oval, fleshy berry.

HISTORY.

Native of Britain, and gives a beautiful appearance to the hedges in the month of June.

MEDICAL VIRTUE.

A conserve of hips is made from this, which is more agreeable than that of the red rose, and is used for the same purpose.

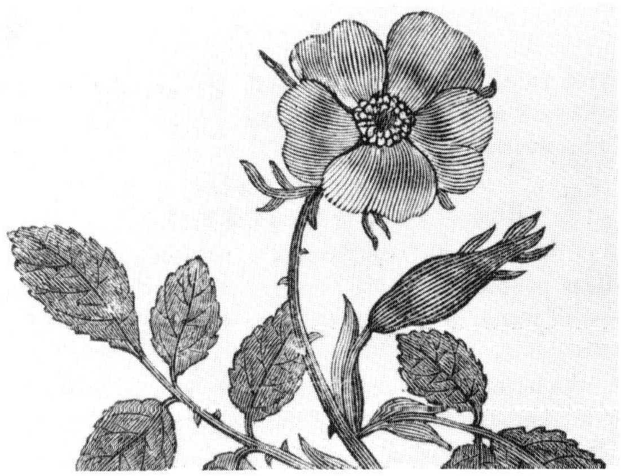

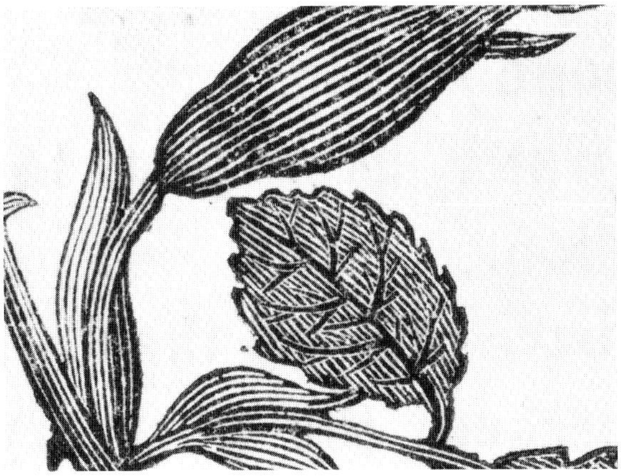

42. DOG ROSE, / ROSA CANINA. Wood engraving, white-line style, engraved by Thomas Bewick (English 1753-1828) after a drawing by Peter Charles Henderson (English fl. ca.1791-1829), and accompanying text. Forming an illustrated page (type area 17 x 8.7 cm.) printed at London in 1810 by Richard Taylor. Printed in black ink on wove paper (top side, page size 28 x 12.7 cm.). In: R. J. Thornton. *A New Family Herbal.* London, R. Phillips, 1810. Page 500. Nissen BBI #1954.

The quality of woodcut botanical illustration declined rapidly towards the end of the 16th century, and for two centuries wood ceased to be a significant medium for printmaking. In the mid-18th century the French engraver Jean-Michel Papillon revived interest in its use and first demonstrated the potential of end-grain boxwood engraving for fine work.

In England it was the genius of Thomas Bewick that gave end-grain wood engraving its great stimulus in the 1780s. His minutely engraved blocks were exercises in consummate craftsmanship, quite apart from their vigor, charm and wit. But they required more than good pressmanship for their successful printing. Bewick was singularly fortunate that contemporary improvements in the manufacture of paper gave printers the smooth surface necessary to do justice to his minute work. "Laid" paper, identified by a regular ribbed pattern when seen against the light, had a similar texture on its wire (i.e. paper mould) side. A delicate engraving sometimes failed to penetrate the troughs of this pattern and gave an unfortunately "ribbed" impression. So-called "wove" paper (made on a mould with a finely woven cloth-like wire mesh) had no such visible surface or see-through pattern. Although invented in the late 1750s, it was not taken up by the printing trade until the end of the century. But that was early enough for Bewick to see all his best work printed to perfection. The introduction of iron presses in the 1800s further improved printing quality.

Apart from starting a tradition for exquisitely minute craftsmanship in wood engraving, Bewick also developed the novel syntactical element of "white line" drawing. The surface of an unengraved block would print as solid black, but any furrow or puncture made on it printed as a white line or dot. Larger white passages could be gouged out, and it was possible to engrave a whole image in white line. Black line was created by cutting white spaces on either side, leaving it standing as a narrow ridge, a process similar to the old woodcut manner. Bewick's particular genius lay in using both types in a design and so created an entirely novel tonal style.

His most famous work was zoological—mammals, birds and people held a fascination for him. His botanical output was limited to providing cuts for this Family Herbal. *His figures have an uninspiring simplicity but show clearly how he drew his subject in black and white lines. His ingenuous visual syntax blends with the directness of the text in a book intended for everyday use in the home.*

the style (with nut suspended below) slides out of the hole, around which the perianth-lobes are connate, until its further progress is arrested by the button-like

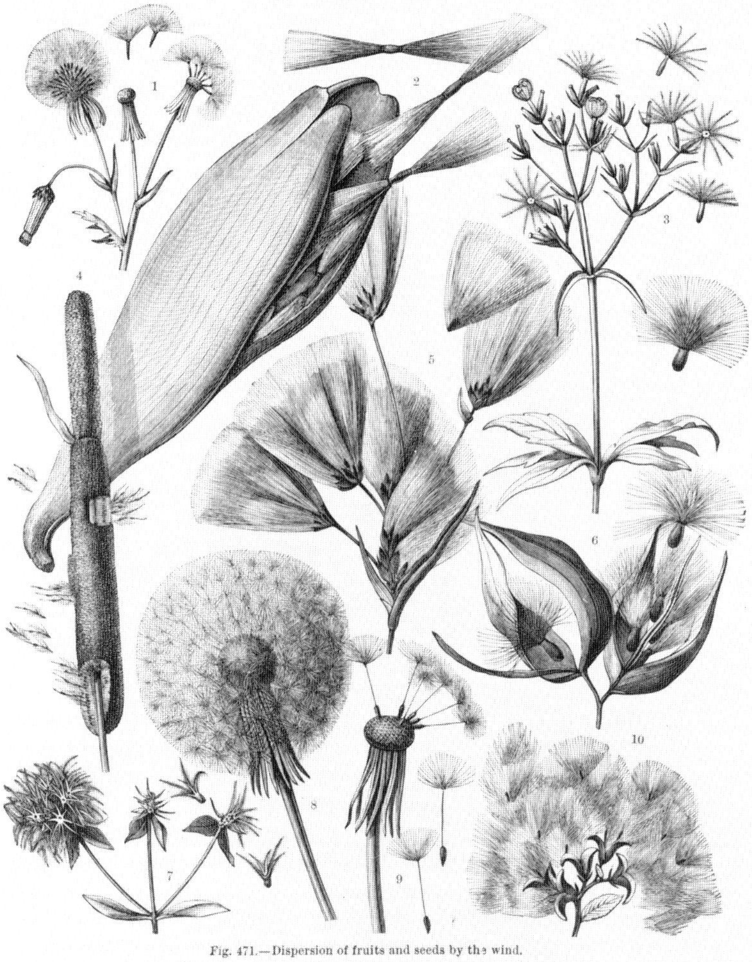

Fig. 471.—Dispersion of fruits and seeds by the wind.

[1] *Senecio vulgaris.* [2] *Adenium Honghel.* [3] *Valeriana tripteris.* [4] *Typha Schuttleworthii.* [5] *Eriophorum angustifolium.* [6] *Cynanchum fuscatum.* [7] *Micromeria nervosa.* [8] and [9] *Taraxacum officincle.* [10] *Salix Myrsinites.*

stigma. The perianth here forms a beautiful parachute, with the nut hanging freely below at the end of a string, like an enterprising balloon-gymnast.

From the fruits and seeds equipped with parachutes we pass to those which are embedded in masses of wool or in envelopes of silky hairs, and are thereby enabled

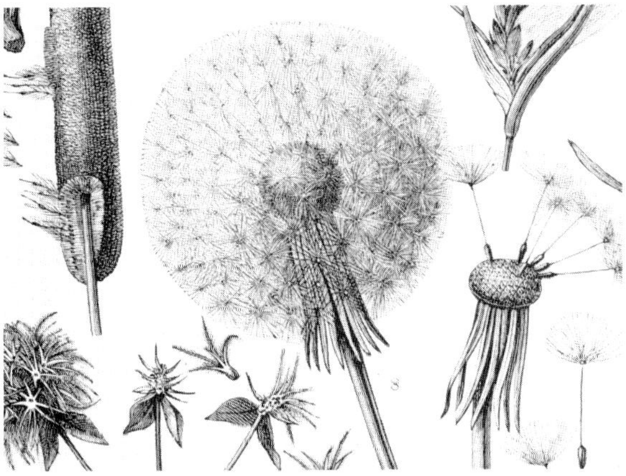

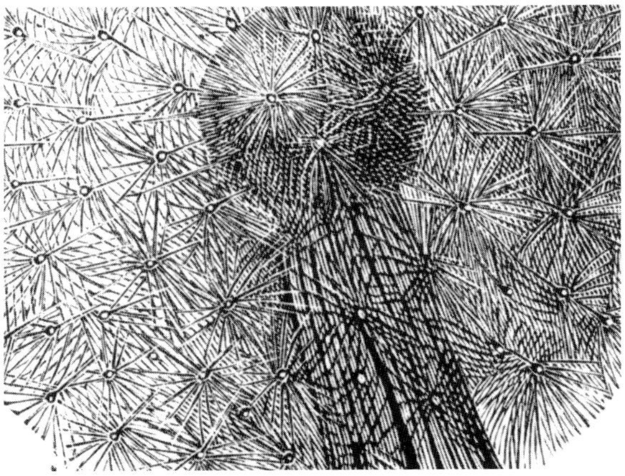

43. Dispersion of fruits and seeds by the wind. Black-line wood engraving, with parallel and cross-hatched shading, artist and engraver unknown. Forming an illustrated page (type area 21 x 13.3 cm.) printed at Glasgow ca.1890 by Blackie & Son Ltd. Printed in black ink (image area 15.8 x 12.9 cm.) on machine-made wove paper (calendered finish, page size 27 x 19.5 cm.). In: A. J. Kerner, Ritter von Marilaun. *The Natural History of Plants.* From the German...by F. W. Oliver. New York, H. Holt & Co., [1895-1896?]. Page 857. Original edition published at Leipzig, 1887-1891. Nissen BBI #1033.

Wood engravings, like the early woodcuts, could be printed at the same operation as the letterpress. That was one reason why the process was so successful for popular illustrated literature. Intaglio processes always required separate printing, often needed special paper, and the prints had to be specially inserted at appropriate places in the book when it was bound.

In Bewick's day, with the iron hand press still in use, his engraved blocks had to be sturdy enough to withstand heavy printing pressure, and lines, though fine, had to be closely laid. The middle and later 19th century saw developments in machine printing, the introduction of smooth calendered paper, and improved ink quality, all of which permitted the bite of 18th century printing to be softened to a "kiss." Finer presswork induced engravers to cut more minute and delicate blocks, and from the 1860s onwards they began to rival copper engraving for fineness of detail.

These qualities are admirably exemplified in the illustrations to Kerner's popular, but scholarly, botanical text book. The work is liberally illustrated, with the engravings generally interspersed throughout the relevant text. The calendered paper has taken clean impressions of blocks that must challenge the limits of fineness. Inking is strong but even and free of any trace of clogging or apparent "squash." The quality and accuracy of these illustrations are readily apparent.

The engraver of these illustrations remains anonymous. He was probably German, but he could have been an Englishman. Bewick's workshop spawned a generation of skilled wood engravers, several of whom went to Germany to establish the craft there. According to one German authority, some 22 English engravers worked at Leipzig from about 1839 onwards, and that is where the original edition of Kerner's book was published in 1887.

This relief-printed medium was probably the most widely used process for general book and magazine illustration for most of the century. But, despite the refinement in engraving and printing seen here, wood engraving never seriously challenged the supremacy of intaglio-engraved or lithographed plates for botanical illustration.

across, white, tinged with violet-purple, expanding in the afternoon and closing on the following morning. D. fastuosa is a handsome species, having white blossoms smaller than the preceding; there is a fine variety of it with the tube of the flower violet and the inside white. The most striking forms of this species humilis flava of the gardens; but although they offer a greater variety of colour, they are less hardy than the older forms just described, and appear to require a warmer climate for their complete development. D. meteloides is a handsome Mexican plant, called in gardens Wright's Datura. Isolated specimens of it have a

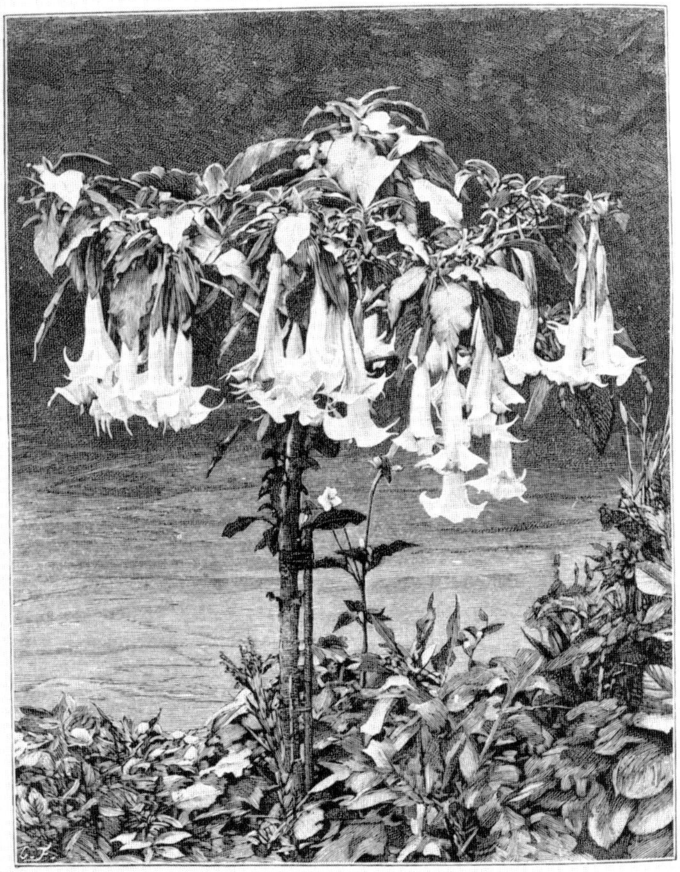

D. cornigera (Brugmansia Knighti) in the flower garden.

bear "double" flowers, the primary corolla having a second and sometimes a third corolla arising from its tube, all being perfectly regular in form, and often being particoloured, as in the single variety with violet flowers. D. fastuosa Huberiana of the seed catalogues, and several varieties of it that are offered, are reputed to be hybrids of this species with the dwarf D. chlorantha flore-pleno or D.

fine aspect in sunny but sheltered nooks. It is from 3 to 4 ft. high, has wide-spreading branches, and blooms from the middle of July till frost sets in, the flowers white, tinged with mauve; from 4 to 6 in. across, showy and sweet, but the leaves emit a disagreeable odour. Besides these there are other kinds in cultivation, such as D. ferox and quercifolia, but those described are the finest. Fresh seeds are

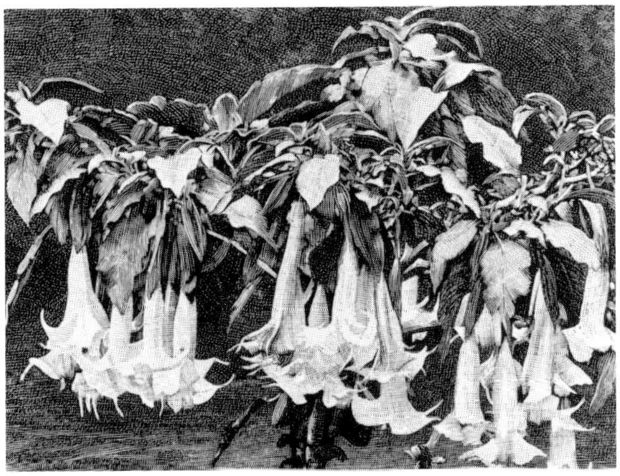

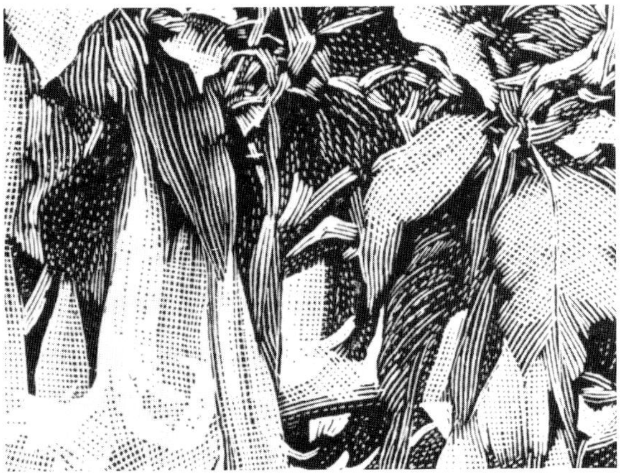

44. D. cornigera (Brugmansia Knighti) in the flower garden. Wood engraving, black-and-white-line "photographic style," engraved by (G.?) Farlet (fl. ca.1885) probably after a photograph. Forming an illustrated page (type area 19.6 x 11.5 cm.) printed at London in1900 by R. Clay & Sons Ltd. Printed in black ink (image area 12.4 x 10 cm.) on machine-made wove paper (calendered finish, page size 23 x 15 cm.). In: W. Robinson. *The English Flower Garden.* London, J. Murray, 1900. Page 518. First edition published 1883. Nissen BBI #1653.

The language of wood engraving has a varied, and in some respects, wider syntax that many intaglio processes. Since a black line is formed subtractively, by cutting the wood away on either side of it, the engraver is free to make his line of whatever character or breadth he chooses, or to leave a large area of black. A white line, on the other hand, is a direct autographic statement, exactly reproducing the handling of the burin, which permits personal expression. Between the extremes of black and white, the engraver can create an infinite variety of lines and tones.

Bewick usually engraved from his own watercolors. His line was generally sturdy, and though much imitated, retained its unique identity. In later years wood engravers were normally employed to reproduce the designs of others, translating their line or tonal drawings into the language of the boxwood medium. By the mid-19th century there was a vogue for the "facsimile style," in which the engravers' skill was directed at copying every caprice of pen-and-ink drawings, thus abandoning their own interpretive skills. Another avenue of change, especially in America, lay in greatly refining and perfecting the tonal elements that Bewick's style had first demonstrated. This concern with the creation of delicately modulated tones and subtle contrasts of light and shade clearly reflected the advent of photography. There was even a trend towards direct imitation of the tonal values of photography and, while early photomechanical processes were still weak and lacking in conviction, wood engravers could often surpass this new technology that threatened to extinquish their craft.

This example, probably based on a photograph, demonstrates the engraver's skill in obtaining these vivid new effects. There is scarcely any trace of black line on the block. A complicated white-line syntax uses skillful lines, dotted lines and flicks to create the form, perspective and lighting. The blooms are treated with a cross-hatched mesh of white line that creates black-dotted tones not unlike those produced in photomechanically screened halftone. The whole is, in fact, more vivid than the average photoengraved illustration of the day, and it gives the reader a very accurate portrayal of the habit and appearance of this decorative Datura.

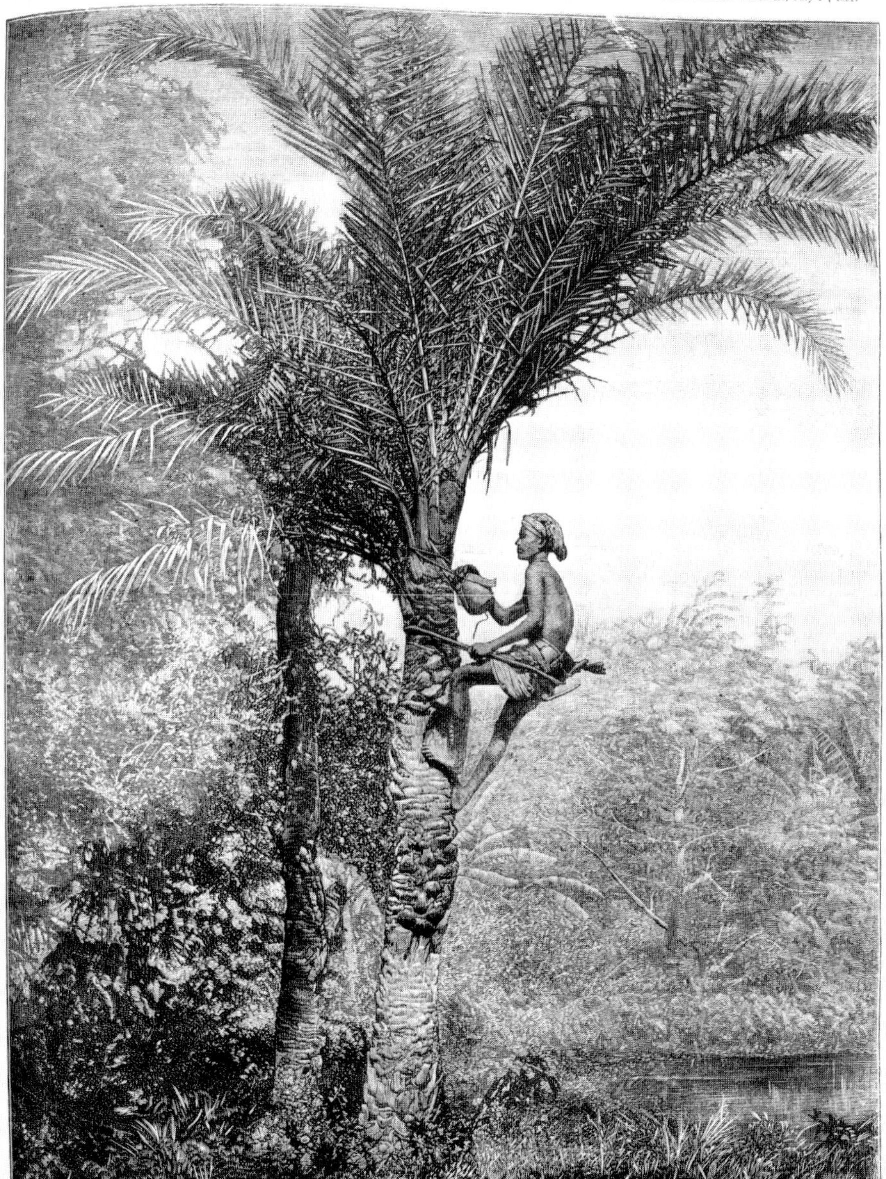

FIG. 14.—PHŒNIX SYLVESTRIS, AN INDIAN "TODDY" PALM: A TODDY-DRAWER AT WORK. (SEE P. 104.)

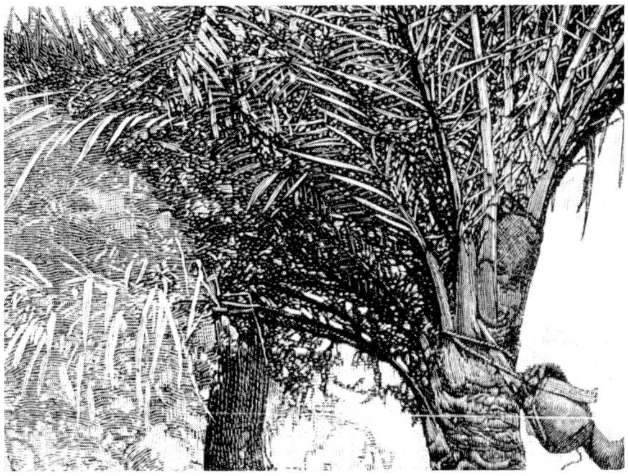

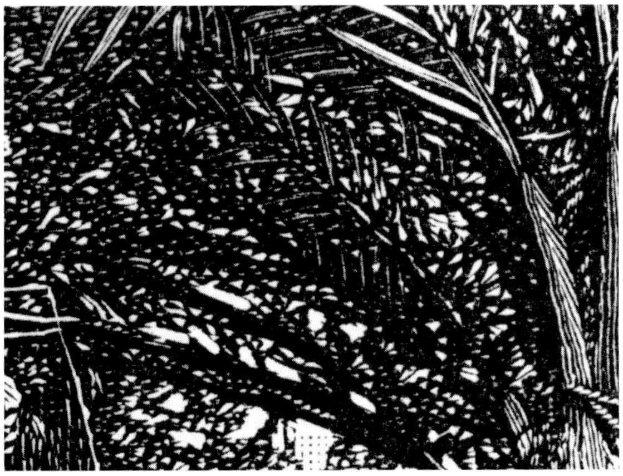

45. PHOENIX SYLVESTRIS, AN INDIAN 'TODDY' PALM: A TODDY-DRAWER AT WORK... Wood engraving, black-and-white-line tonal style, with parallel and cross-hatched shading, engraved by Worthington George Smith (English 1835-1917) after a photograph by Johnson & Hoffman of Calcutta. Forming a full-page illustration printed at London in 1891 by Bradbury, Agnew & Co. Ltd. Printed in black ink (image area 25.5 x 19 cm.) on machine-made wove paper (matte finish, page size 30.5 x 23 cm.). In: *Gardener's Chronicle, series 3.* [Edited by M. T. Masters.] London. Volume 10, 1891. Page 105.

This large engraving exemplifies the extremes to which commercial wood engraving was prepared to go in imitation of photography. At a distance it assumes the look of a photographic halftone, and indeed the author of the accompanying article praises it as "an admirable reproduction of [the] photograph." It aptly demonstrates the extraordinary imitative skill possessed by later 19th-century wood engravers, in this case by W. G. Smith, long associated with the Gardener's Chronicle, *who was no mean botanist and archeologist, quite apart from his skill with the graver.*

Smith could have had no lofty conceit about creative engraving here. It simply showed readers of the Gardener's Chronicle *the growth habit of this common Indian cultivated palm and the method of harvesting Toddy. Most copies of this weekly magazine were probably soon discarded and the contents very likely forgotten. We many surmise that the original photograph was unsuitable for reproduction by the photomechanical halftone process of the day, perhaps owing to lack of contrast or definition. Smith's engraving is, then, no more than highly competent hack-work but it seems a pretty impressive exercise in technique to our modern eyes, and a lot of effort for an ephemeral, if attractive, magazine illustration.*

It appears that the printer assisted with the modulation of tonal values. Few woodblocks ever printed entirely satisfactorily without the intervention of the pressman's skill. To ensure a clean impression on the cylinder machines of those days, the pressman had to make adjustments to the local pressure applied to densely or sparsely engraved parts of the image. By a process of trial and error, the impression roller, which forced the paper onto the inked block, was "made ready" with overlaid slips of thick or thin paper to "bring up" certain passages in the impression. Smith's block appears to have been treated in this way. The change in value between the left foreground and background, and the "forwardness" of the whole plant, were probably created by the pressman's "make-ready."

Boxwood blocks came in very small sizes. The average tree girth allowed a maximum of about five by five inches, but many were much smaller. A block of this size was made by fastening together several smaller blocks. Unfortunately, boxwood was liable to radial shrinkage, and this print shows a distinct horizontal separation of the blocks across the center. By this date the usual practice was to make electrotyped metal copies of wood blocks for working on the press, which would have prevented the defect seen here.

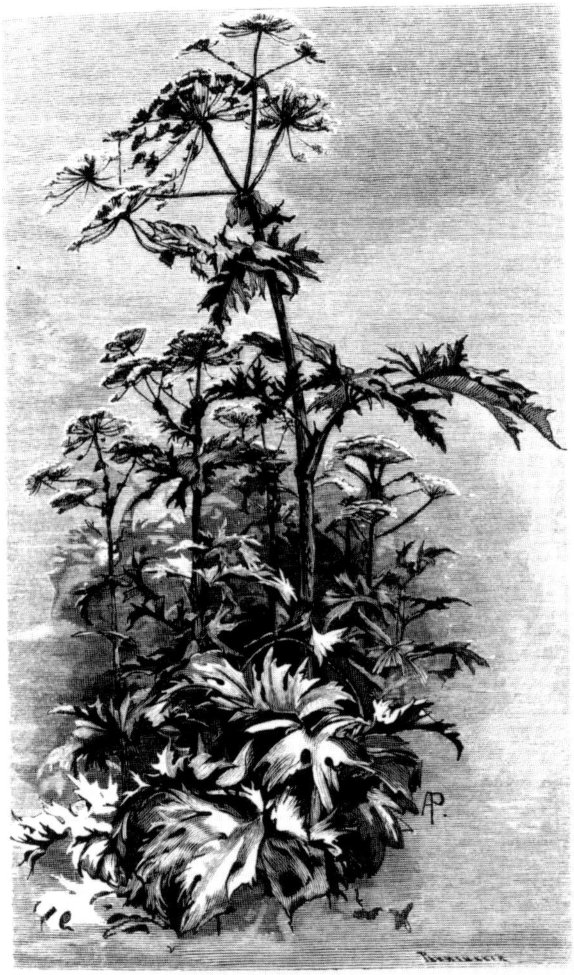

GIANT COW PARSNIP. Type of Great Siberian herbaceous vegetation.
For rough places only.

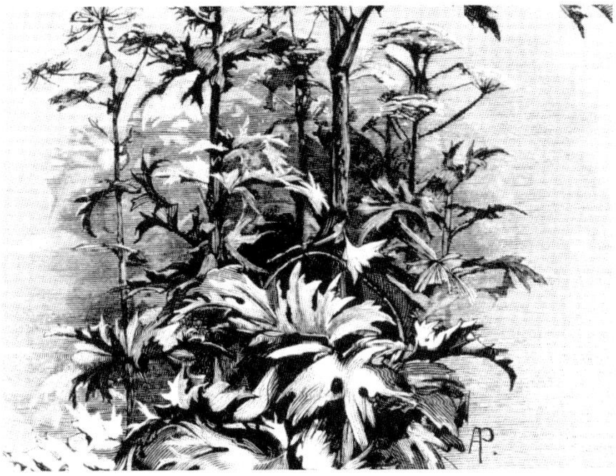

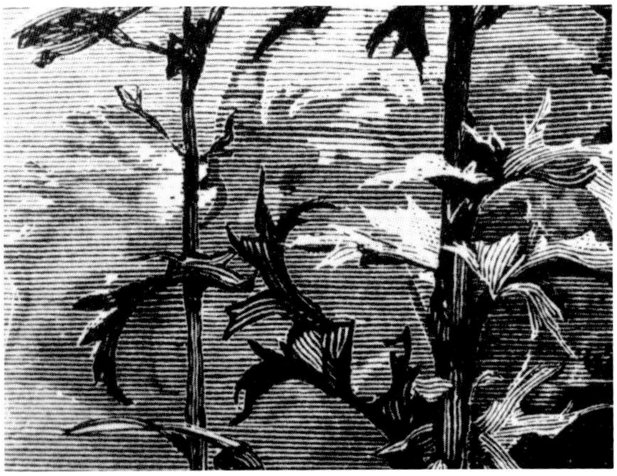

46. Giant Cow Parsnip. Wood engraving, white-and-black-line style, engraved by Stephane Pannemaker (Belgian 1847-1930) after a drawing by Alfred William Parsons (English 1847-1920). Forming a full-page illustration printed at Oxford in 1895 by Horace Hart, printer to Oxford University. Printed in black ink (image area 13 x 8 cm.) on machine-made brilliant white wove paper (top side, calendered finish, page size 21.7 x 14 cm.). In: W. Robinson. *The Wild Garden.* [Fourth edition.] London, J. Murray, 1895. Page 45. First illustrated edition published London, 1881.

By the closing years of the 19th century, wood engraving had almost completely given way to photomechanical engraving. Relief-engraved halftones and line blocks could reproduce the artist's original work and, like wood engraving, be printed with the letterpress. Not everyone, though, was happy with this trend—certainly not William Robinson, the horticultural writer. In the preface to the new edition of his Wild Garden *he remarks, "I am happy to be able to illustrate the book with good wood engravings in these days of many 'processes,' often called 'improvements,' but which, so far, are its ruin. The few cuts done in the former edition by such processes have been re-engraved on wood for this." He had employed Alfred Parsons to prepare drawings "after nature" to illustrate his ideas on this style of gardening and the most suitable kinds of plants.*

Parsons' drawings are engagingly decorative, sometimes quite ethereal, and despite being self-consciously "artistic" are generally informative. They have been superbly engraved, in this case by Stéphane Pannemaker, better known as an interpreter of Gustave Doré's work.

But there are no efforts at photographic effects here. Instead, the engraver made a technically admirable interpretation of Parsons' drawing. White- and black-line engraving provided a masterly representation of the vivid highlights and shades that Parsons used to emphasize the wild grandeur of this great hogweed. The whole personality of the plant is highlighted by its depiction against a curiously ethereal middle-tinted ground lacking any apparent perspective. It was unusual, by this time, to apply such creative engraving to a plant illustration.

Robinson had the book printed on brilliant paper that imitated the quality of the best handmade. Elsewhere, he acidly expressed his displeasure with the "tin-shine paper" that contemporary halftone illustrations required for successful reproduction. This was his emphatic preference and it also served to throw his beloved wood engravings into theatrical focus. They were among the last products of the 19th-century tradition for minutely fine wood engraving that Bewick had founded in the 1780's. Botanical illustration parted company with wood engraving at about this time. In the present century, wood engraving was to find a new personality through its connection with the private-press movement.

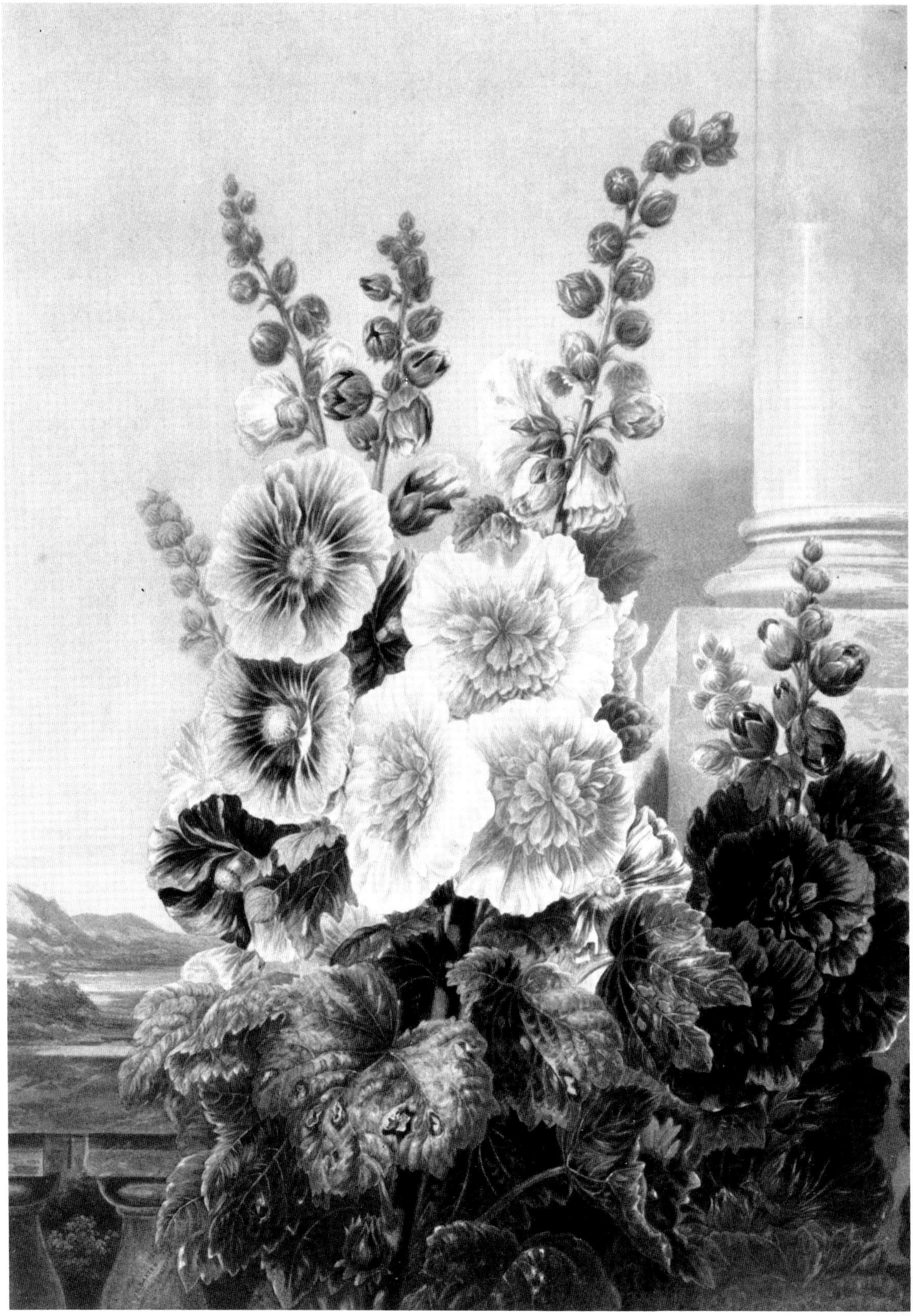

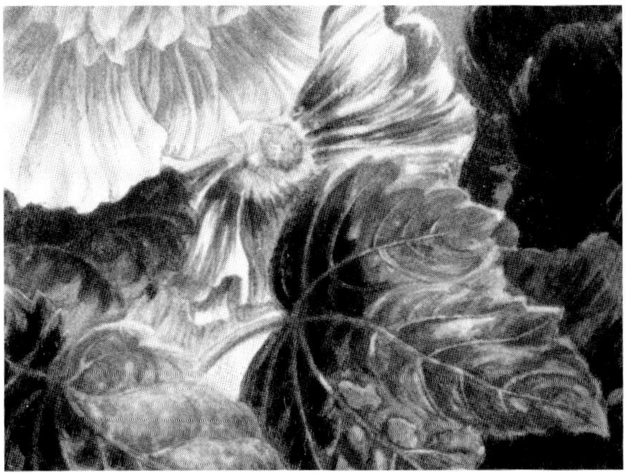

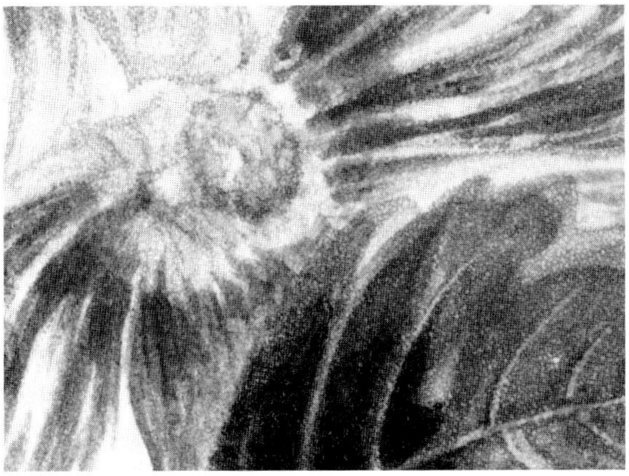

47. HOLLYHOCKS / FROM A PAINTING BY / V. BARTHOLOMEW ESQRE. Aquatint-etched key plate, with several wood-engraved tint blocks, the key probably engraved by Joshua Gleadah (English fl. 1816-1874) and the tint blocks by George Baxter (English 1804-1867) after a painting by William Valentine Bartholomew (English 1799-1879). A separately published picture printed and published at London in 1857 by George Baxter. Color-printed by the "Baxter process" (black-printed intaglio "key" (image area 38 x 27 cm.) overprinted with 11 "oil-colors") on machine-made wove paper (laid down on two-ply board, sheet size 49 x 37.2 cm.).

Soon after Bewick reestablished wood engraving as a major graphic process in Britain, experiments were directed toward adapting it to color printing. None of these early experiments had any relevance to botanical printmaking until the arrival of George Baxter's process, patented in London in 1835. Baxter was a painstaking craftsman with particularly high standards of presswork. His "invention" involved nothing more original than overprinting an intaglio-engraved "key" print with several colored tints from wood-engraved blocks. Rather, his claim to originality lay in the elaborate craftsmanship of his impeccably delicate "key" plates, his precisely cut tint blocks, his sensitive choice of colors, and above all his superb pressmanship. With these skills he produced prints that were richly eyecatching and susceptible of the closest scrutiny. Most of his best work was published as separate pictures, but he also prepared a considerable number of smaller prints for book illustrations.

This example is typical of the rich effects he achieved. The monochrome key plate is a superbly aquatinted engraving that could stand as a print in its own right. It has all the line, shading and tinting necessary for the correct modelling of the whole. The relief overprinting has added a full palette of 11 transparent colors to recreate the appearance of Bartholomew's original painting. Baxter rarely printed colors over each other; he preferred to cut a separate block for each color and print them all with needle-fine register.

Baxter always boasted of using "oil colors," and the glossiness of the inks can be perceived readily in the print. His colors were concocted from vegetable, rather than mineral, pigments, which may account for the warmth and freshness he often achieved. However, they rapidly fade when exposed to sunlight. A contemporary Italian color maker pointed this out to Baxter, offering mineral colors instead, but the printer, always pigheaded, refused to take heed.

The most extensive and successful application of the Baxter process to botanical illustration was made by one of his patent licensees, William Dickes, who in the 1850s printed over 300 delicate plates for Anne Pratt's Flowering Plants & Ferns of Great Britain.

Valentine Bartholomew, whose painting is represented here, claimed the title of "Flower Painter in Ordinary to the Queen" [Victoria] and, according to Blunt, was probably the last artist to hold such office.

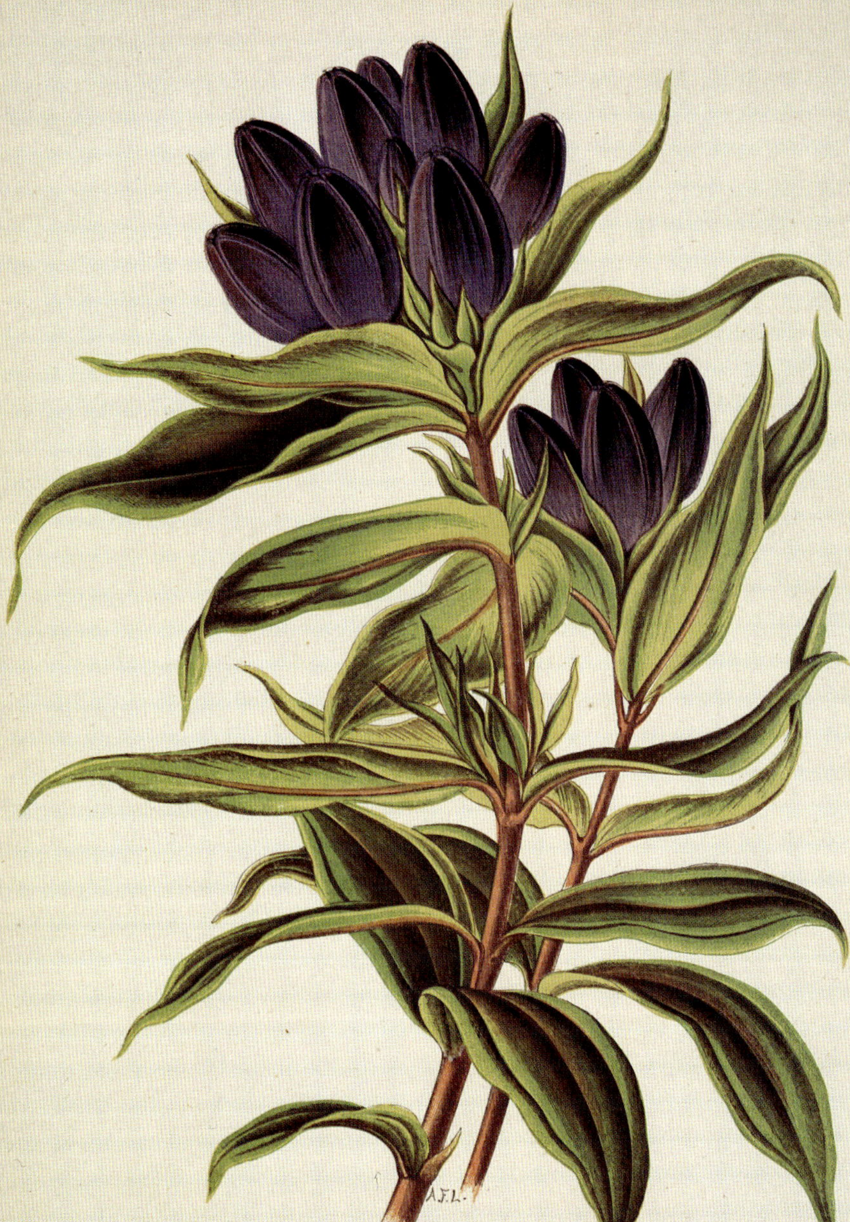

GENTIANA ANDREWSII.

XV

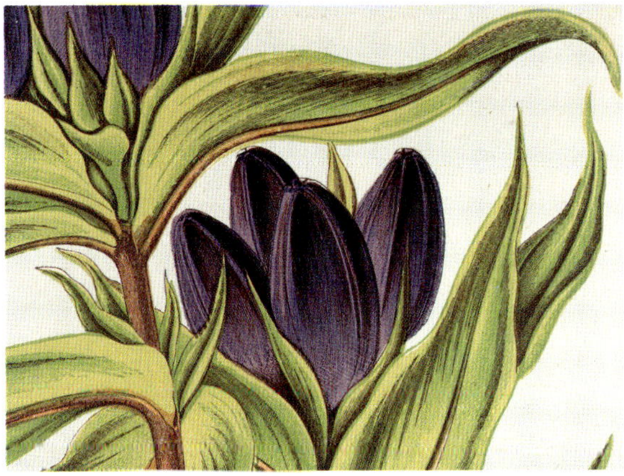

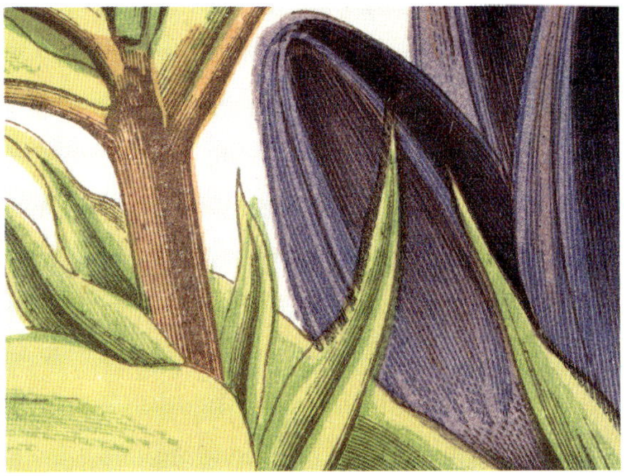

48. GENTIANA ANDREWSII. Outline wood engraving, with parallel-line shading, on six color-separated blocks, engraved by Alexander Francis Lydon (English 1836-1917), artist unknown. Forming a full-page illustration printed at Driffield, Yorks. ca.1871 by Benjamin Fawcett. Color-printed, the "key" in black, the five tint blocks in light and dark green, light and dark mauve, and russet inks (image area 20.5 x 14 cm.) on machine-made wove paper (matte finish, page size 25.3 x 17.5 cm.). In: D. Wooster. *Alpine Plants.* London, G. Bell & Sons, [1871-]1872-1874. Plate 15. Nissen BBI #2186.

A number of British printers used wood-engraved relief blocks as the means of creating color-printed pictures, but only those who had paid Baxter's high license fee were permitted to use his process. The patent expired in 1854, but by then several printers had mastered the technique of high-quality relief color printing, or chromoxylography, without the need for an intaglio key plate. A Baxter print required both copper-plate and relief processes for its production, so it was cheaper and simpler to dispense with the intaglio anyway.

One of the most successful chromoxylographic printers, especially in natural-history illustration, was a skillful provincial printer, Benjamin Fawcett. He owed much to the extraordinary industry and artistic talent of his employee Frank Lydon, who drew and engraved this illustration, one of 108 prepared for the Alpine Plants.

Lydon engraved a strong outline and some sparing parallel-line shading on a black-printed "key" block. Much of the modelling of the whole image was provided by the tint blocks. He made five separations to add color to the "key," but they were more than simple flat tints; each included line shading to provide modelling and tonal variation. Baxter had recognized the essential importance of accurate register for convincing color printing, and Fawcett rarely failed on that score either. His hundreds of color-printed illustrations maintained a uniformly high standard with attractively bright, and usually quite accurate, colors.

Apart from the lack of an intaglio key plate, what specially distinguished this work from Baxter's was the quality of the colored inks. Fawcett's had no oily shine. He preferred mineral pigments and an ink varnish with a matte finish. We have no information on the composition of his inks, but Fawcett's printing has exactly the same "dry" finish as work done by William Savage, a distinguished early 19th-century English color-block printer. Savage successfully avoided the oily staining that disfigured some earlier color printers' work by using copaiba balsam for mixing his ink. Savage described his formulae in a book on ink manufacture and Fawcett may well have chosen to use them.

Cèdrus Libàni.

The Cedar of Lebanon.

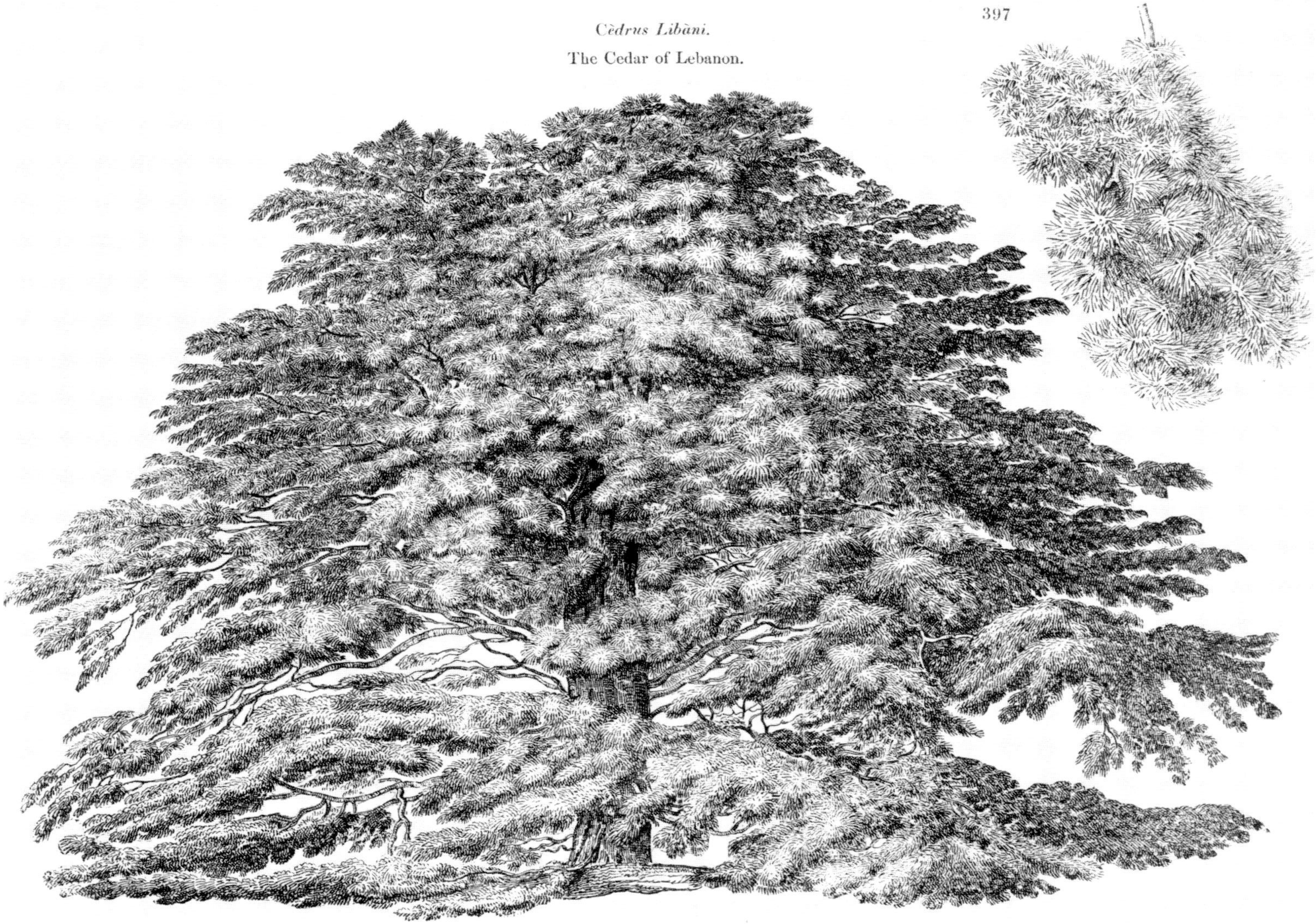

Full-grown tree at Syon, 72 ft. high ; diam. of the trunk 8 ft., and of the head 117 ft.
[Scale 1 in. to 12 ft.]

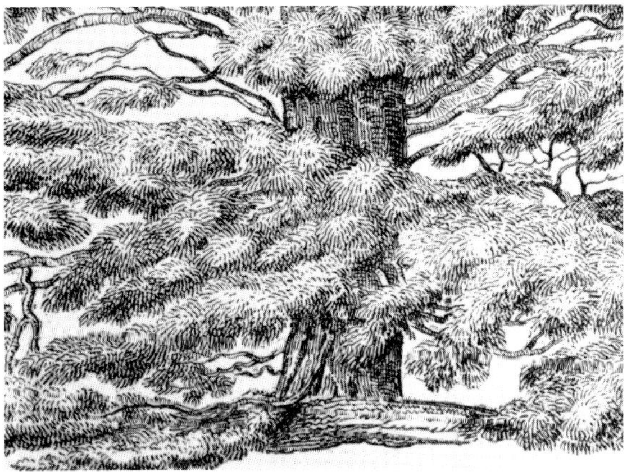

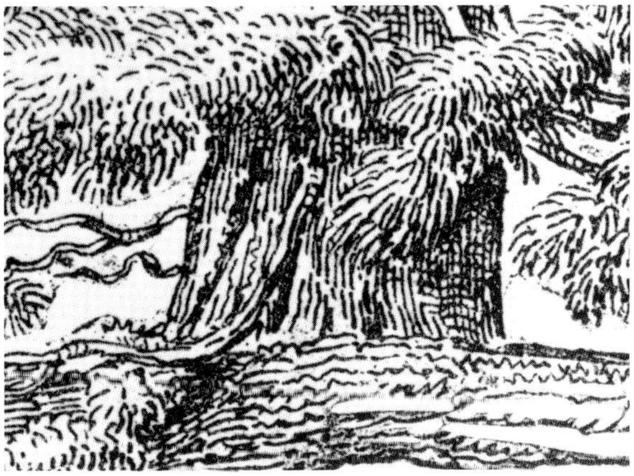

49. Cidrus Libani. The Cedar of Lebanon. Relief block, the cast of a deeply bitten line etching (Branston's relief process), etched and finished by Robert Edward Branston (English fl. 1827-1885) after a drawing by G. R. Lewis (English fl. 1830s). Forming a double-page illustration printed at London in 1844 by Andrew Spottiswoode. Printed in black ink (image area 17.5 x 24.5 cm.) on machine-made wove paper (calendered finish, double-page size 22.5 x 28.5 cm.). In: J. C. Loudon. *Arboretum et Fruticetum Britannicum.* Second edition. London, Longman, 1844. Plate 397. First edition published 1835-1838. Nissen BBI #1238.

The tremendous botanical activity of the post-Linnaean period, coupled with the extraordinary growth in public interest, led to a need for synthetic presentations of this expanding realm of knowledge and economical means for their publication. Existing picture-printing methods were seriously put to the test to meet the rapidly growing demand for illustrated books.

Wood engraving emerged as the only process that could be printed on the steam-driven machines that were so widely introduced into Britain in the 1830s and 40s. Bewick's methods were essentially slow in both engraving and printing. The adaptation of the medium to mass production led to a huge increase in the numbers of wood engravers, but each still worked with painstaking slowness. However, as the relief-block illustrations were so successfully printed on fast machines, efforts became directed at finding substitutes for the engraver's tediously slow handwork. It was an age of inventors and universal technological change, and picture printing had its full share of experimenters.

In the late 1830s, Robert Branston, a leading London wood engraver, devised a moderately successful relief-etched process, intended to replace the very craftsmanship in which he excelled. Quite simply, he made a line etching of his subject, taking care not to lay the lines too close. The copper was etched to an unusual depth and a cast of the image was taken in type metal. The result was a block with the originally etched lines standing up in low relief. The relief image was then lightly ground to make it perfectly level, permitting it to be printed just like a wood engraving. Unfortunately, it was often still necessary to go over the casting with a small gouge to deepen the spaces between some lines so that they would not clog with ink.

J. C. Loudon was one of several compilers of encyclopaedic presentations of the great quantities of new botanical and horticultural data. He had several stout, well illustrated tomes to his credit but his greatest undertaking was the encyclopaedic Arboretum et Fruticetum Britannicum, *for which he gathered nearly 3000 drawings. He used Branston's method for the 412 large tree portraits and also for many of the 2546 text figures. This is a fine example of the process and apparently free from any hand finishing.*

Although the process seems to have worked well for this book it found little success elsewhere. Loudon suffered a crippling financial loss from this publication. Perhaps the technical success of Branston's process was not matched by savings in cost. It appears to have been well suited to this type of illustration, the delicacy of line allowing depiction of complicated detail that would have been very taxing to engrave on wood.

ORDER CIX. PANGIACEÆ.—PANGIADS.

Pangiaceæ, *Blume in Ann. Sc. N. ser.* 2. 88. (1834); *Bennett in Horsfield, pl. Javan. p.* 208; *Endl. Gen. p.* 922.

DIAGNOSIS.—*Papayal Exogens, with polypetalous flowers, which have scales in the throat, of the females.*

Trees. Leaves alternate, stalked, entire, or somewhat lobed. Flowers axillary, solitary or fascicled, or in few-flowered racemes, ♂ ♀. Sepals 5, rarely 2, 3, or 4. Petals 5, rarely 6. Scales as many, opposite the petals. ♂ Stamens 5 or 00; not a rudiment of ♀. ♀ Sterile stamens equal in number to the petals, rarely more. Ovary free, 1-celled; ovules 00, attached to 2- 6 parietal placentæ. Capsules succulent, indehiscent, 1-celled. Seeds 00, large; albumen abundant, oily; embryo nearly as large as the albumen; radicle protruded; cotyledons generally leafy and veined. —*Bennett.*

What the distinction is between these plants and Papayads, except that the last are monopetalous, and have no faucial scales in the ♀ flowers, it is hard to say. Mr. Bennett throws no light upon the matter, and I am unable to supply any.

The species are found in the hotter parts of India. All are poisonous plants. The seeds of Gynocardia odorata are employed extensively by the natives of India in the cure of cutaneous disorders. When freed from the integuments they are beaten up with clarified butter, into a soft mass, and in this state applied thrice a day to the parts affected. Chaulmoogra and Petarkura are the native names. Hydnocarpus venenatus, a Cingalese plant related to this Order, bears a poisonous fruit, which when eaten occasions giddiness and dangerous intoxication. It is used in Ceylon for poisoning fish, which afterwards become so unwholesome as to be unfit for food. According to Rumphius, the plant Pangi has a hard, solid wood, and its kernels are boiled, then cut to pieces and macerated in cold water to remove the noxious narcotic qualities;

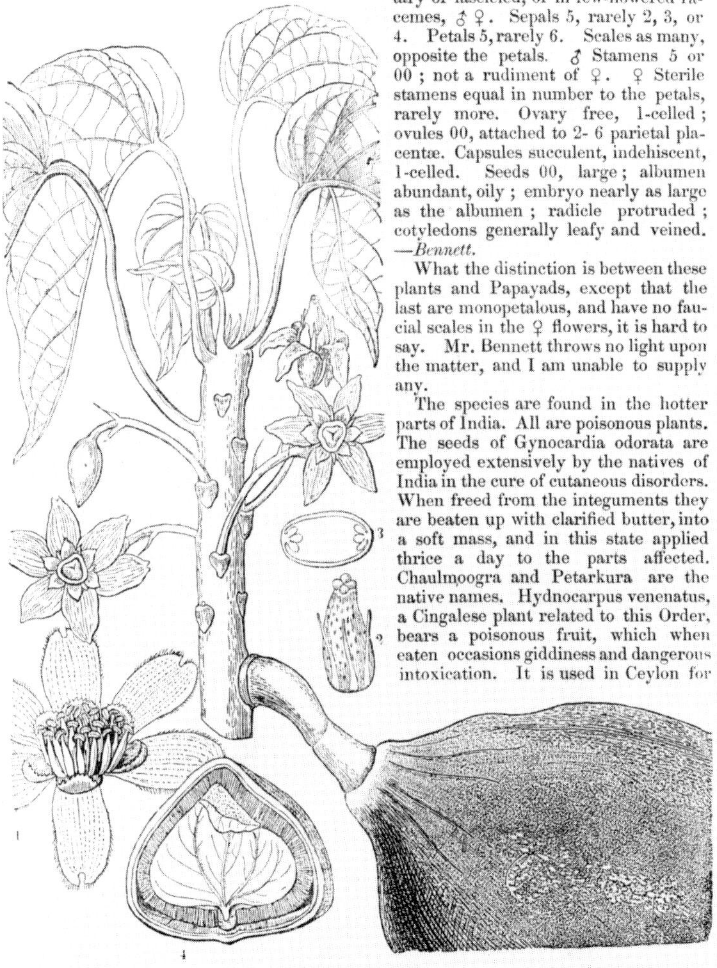

Fig. CCXXIII.

Fig. CCXXIII.—Pangium edule.—*Horsfield.* 1. a ♂ flower; 2. an ovary with its sterile stamens; 3. a cross section of an ovary; 4. a perpendicular section of a seed.

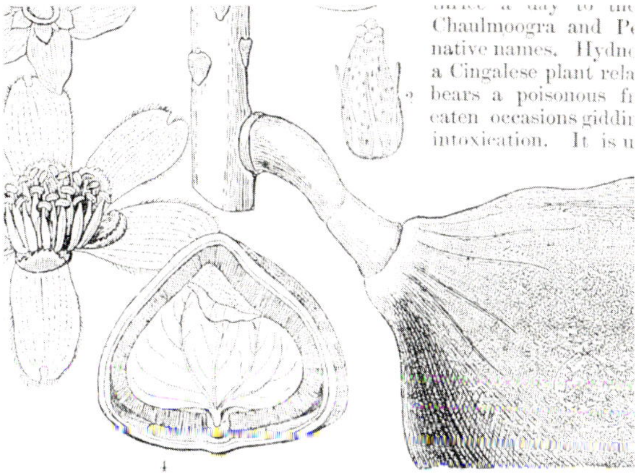

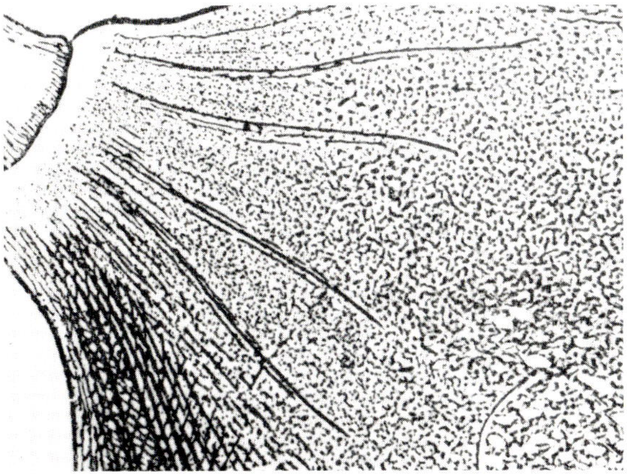

50. Pangium edule.—Horsfield. Relief block, cast from a line and cross-hatched drawing incised into a "composition" applied to a metal plate (glyphography), engraver and artist unknown, with accompanying text. Forming an illustrated page (type area 18.5 x 10 cm.) printed at London in 1853 by Bradbury & Evans. Printed in black ink (image area 13 x 10 cm.) on machine-made wove paper (matte finish, page size 22.5 x 14.5 cm.). In: J. Lindley. *The Vegetable Kingdom.* Third edition. London, Bradbury & Evans, 1853. Page 323. First edition published 1846.

The singular merit of Branston's process was that it had the same autographic advantages as intaglio etching. The lines drawn by the artist were the lines of the relief image. Wood engraving operated in reverse. The artist's drawing, usually made direct on the surface of the block, either had each line cut around with extraordinary care or, if done in wash, had to be "interpreted" by the engraver in his own visual language. Not long after Branston's invention was tried out, a much more successful process was launched under the trade name Glyphography. Like Branston's process, this used a relief-printed autographic medium and was, coincidentally, employed to illustrate another encyclopaedic botanical work.

The process was operated with various modifications in technique, but its basic principle was simple. A copper plate was thinly coated with a wax-based composition and the design was drawn onto it with an etching needle, taking care to incise through to the metal. The "islands" between the lines were built up with additional wax to give greater depth to the incised lines. A cast of the drawing was taken in plaster for stereotyping or an electrotyped copy was made; the end product in either case was a relief metal block. Glyphographs, like wood engravings, could be set and printed with the letterpress.

This example is one of many made in 1845 for the Vegetable Kingdom. *Superficially, it has the appearance of a fine wood engraving, but closer inspection reveals a fundamental difference. The fineness and complexity of some passages are beyond the skill of any wood engraver, and, more significantly, it is clearly a positively drawn image, quite different in character from wood-engraved line.*

Glyphraphy was used with reasonable success for about 15 years, but it never displaced wood engraving. One reason was that, as a patent process, its use was restricted; wood engraving, however, was universally accessible. The three editions of this book were apparently the only application of Glyphography to botanical illustration.

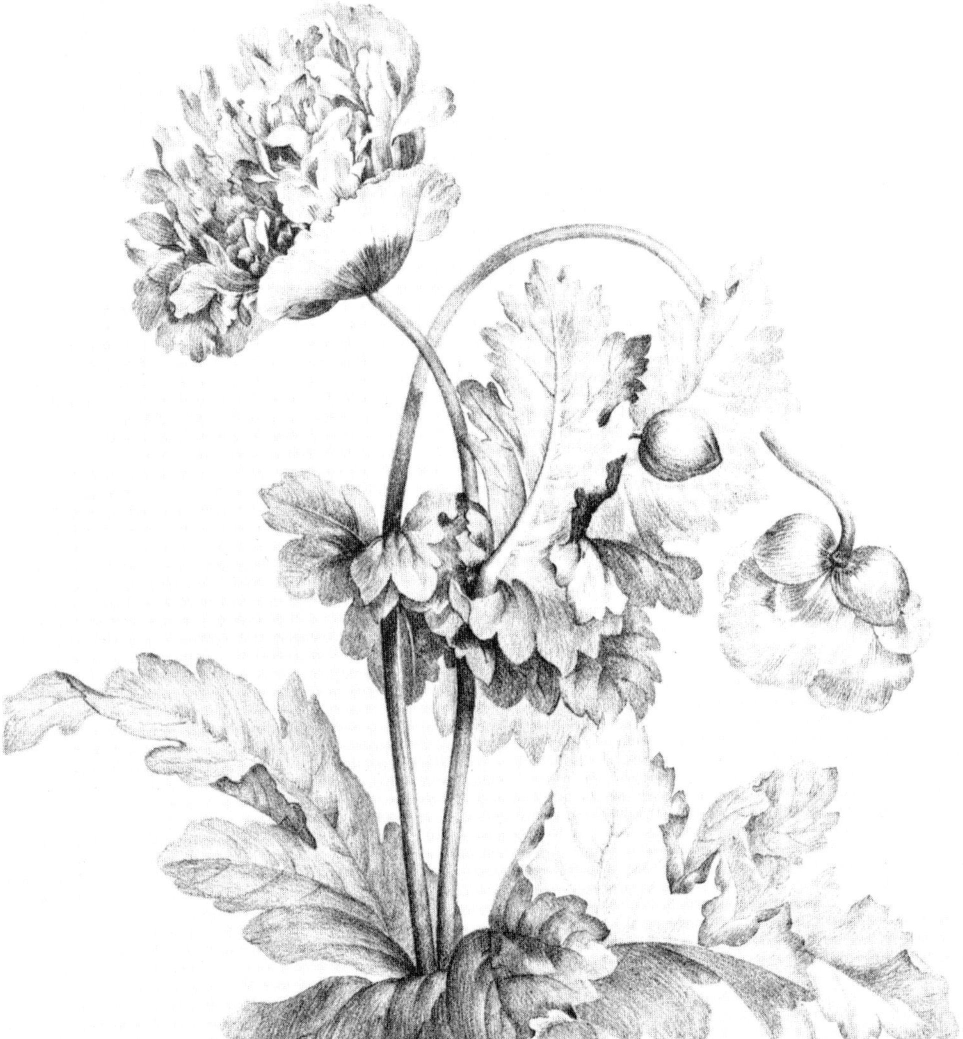

Papaver somniferum. Schlafmachender Mohn

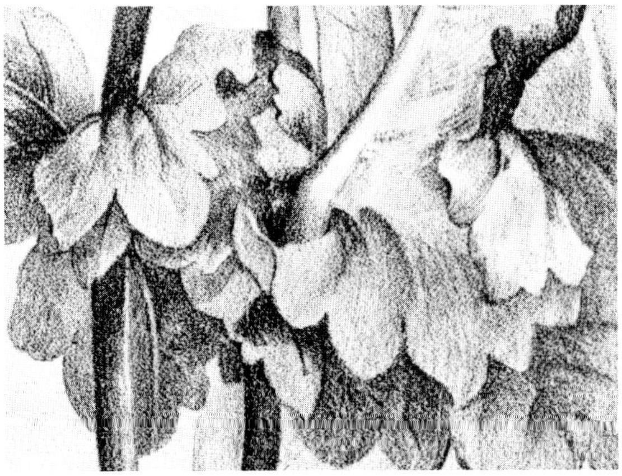

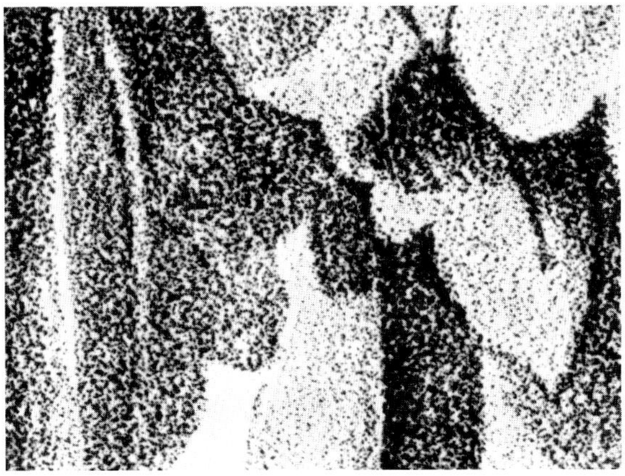

51. Papaver somniferum. Lithograph, chalk style, drawn on stone by Johann Nepomuk Mayerhoffer (Austrian 1764-1832) after his own original. Forming a full-page illustration printed at Munich ca.1811 by Lithographische Kunstanstalt an der Feiertagsschule für Künstler und Techniker (under the direction of Herman Mitterer). Printed in black ink (image ink 34 x 24 cm.) on wove paper (wire side, watermarked "MT. F?A. HUBAR.," sheet size 47 x 32 cm.). Ex: F. von P. von Schrank. *Flora Monacensis.* Munich, Inst. Lith. Scholae Festivalis, 1811-1818. Plate [?]. Nissen BBI #1803.

The accidental discovery of the lithographic process in 1798 and its development into a major printmaking medium was to have much significance for botanical illustration. The immiscibility of grease and water must have been familiar to all since the stone ages. The fortuitous discovery that this could form the basis for a process of "chemical printing" fell to Alois Senefelder of Munich. Using a slab of porous limestone, he related in his treatise on lithography, his process was "to wash the polished stone with soap-water, to dry it well, to write or draw upon it with the composition ink of soap and wax, then to etch it with aqua-fortis; and, lastly, to prepare it for printing with an infusion of gum-water." A roller charged with greasy ink passed over this prepared stone left a deposit of ink on the drawn lines; the wetted parts of the stone were unaffected. When the stone was put through the press the image printed onto paper. This was a startlingly novel addition to the range of printmaking processes.

Lithography found its first major botanical application in 1811, appropriately enough in a flora of Munich, illustrated with no less than 400 plates. This spirited plate of the opium poppy exemplifies the character of an early chalk-style lithograph. It has the look of a soft pencil or chalk drawing done on well textured paper and shares the syntax of tones and shades used by the ordinary draughtsman. Tones could range from solid black to palest grey, but fine linear detail was sacrificed for this "chalk" effect. In consequence the artist was unable to show minute structures, such as the prominent hairs that normally clothe the stem of this poppy.

This was a great revolution in graphic reproduction. Although there is an obvious similarity to the appearance of Roubillac's engraved bouquet (item #22), there was a fundamental difference in the means of their creation. Mayerhoffer's work was the first ripple in a tide of many thousands of botanical lithographs produced during the century.

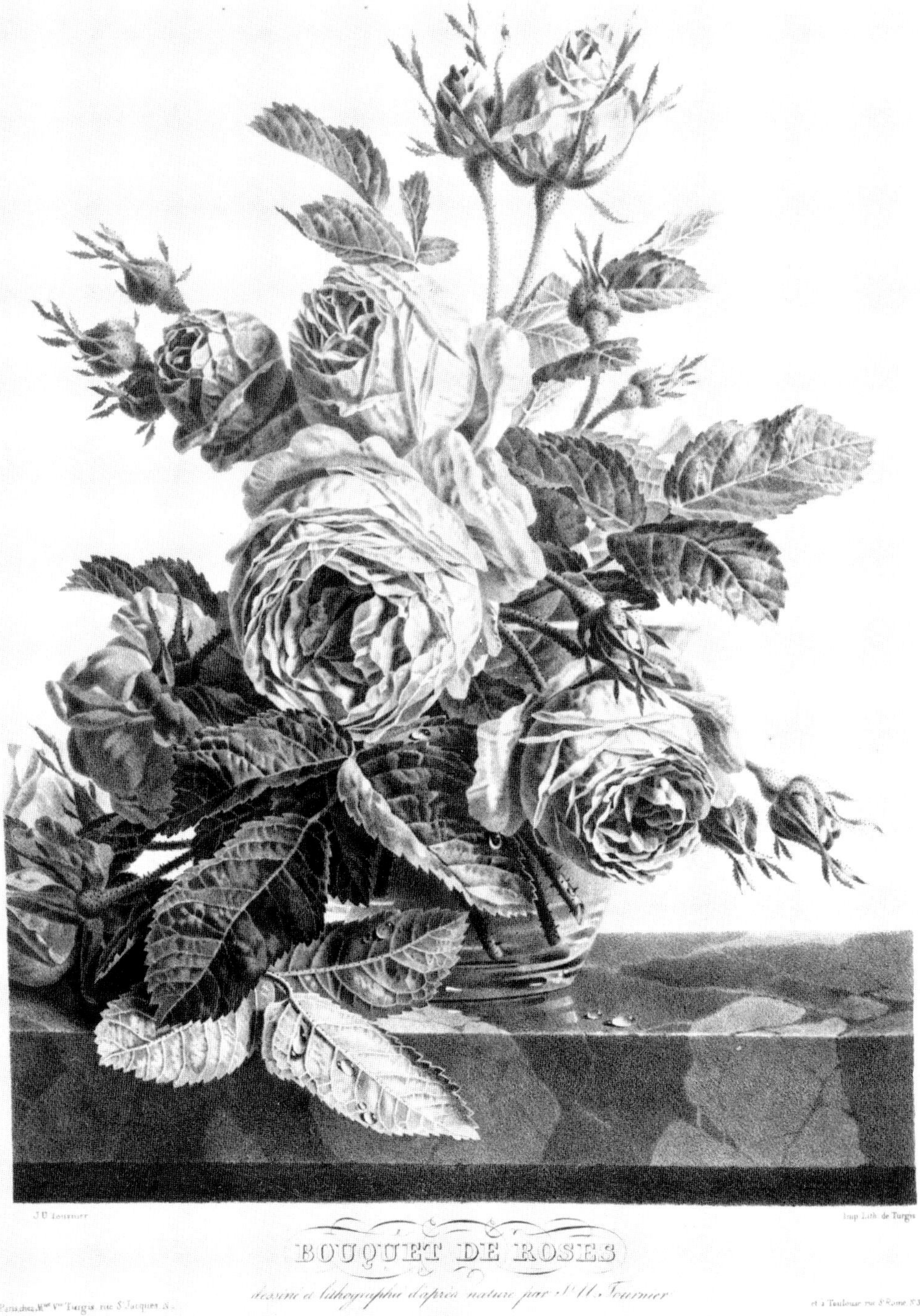

J U Fournier

Imp. Lith. de Turgis

BOUQUET DE ROSES

dessiné et lithographié d'après nature par J. U. Fournier

Paris, chez Mme Vve Turgis rue St Jacques, 8.

et à Toulouse rue St Rome No 30

No 3.

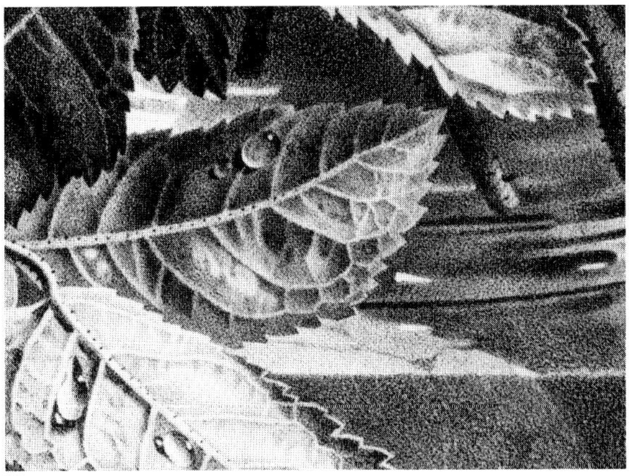

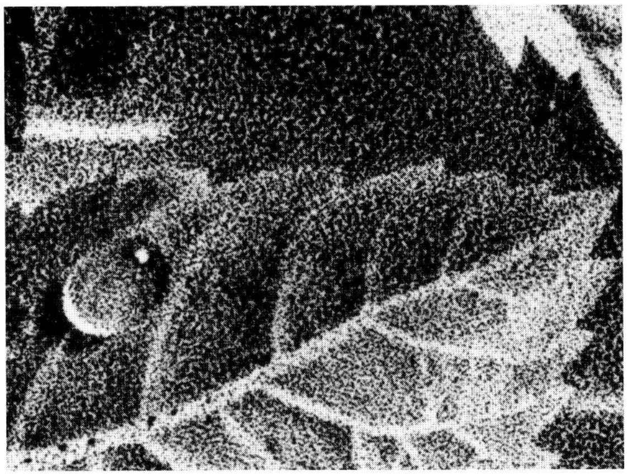

52. BOUQUET DE ROSES dessiné et lithographié d'àpres nature par J. U. Tournier. Lithograph, chalk style, drawn on stone by Jean Ulrich Tournier (French 1802-1865) after his own original. A separately published picture printed and published at Paris in the early 19th century by V. Turgis. Printed in black ink (image area 34.5 x 24.5 cm.) on wove paper (top side, sheet size 41.5 x 28.6 cm.).

The apparent coarseness of the paper texture in Mayerhoffer's pioneer lithograph is somewhat misleading. The paper has a fairly coarse surface texture but even had it been quite smooth the printed image would still have had a coarse-grained appearance. When grinding the surface of the stone, preparatory to drawing, it was possible to create fine, medium or harsh textures by using different grades of abrasive (usually sharp sand). It was the "tooth" of the stone's surface that lent the granular effect to a lithographic drawing, not the texture of paper on which it was printed.

Tournier's fine bouquet of roses was drawn on very finely textured stone which gave the print a very refined character. The early decades of the 19th century saw tremendous improvements in the craft of drawing on stone and in the printing of lithographs. Chalk-style drawings, with their range of diverse tones, had always been particularly demanding of the printer's skill but, as this example shows, high standards were achieved.

The same basic elements exist in both Mayerhoffer's and Tournier's chalk-style illustrations, which were conceived as tonal images, virtually free of line, and with values extending from near white to deep black. However, the fine stone texture, better-quality lithographic chalk, and an increased sophistication in draughting technique combined to provide Tournier with the means of communicating graphic information in much more detail. Every little prickle was drawn, leaf texture was depicted in life-like detail, every petal was carefully delineated, and, with a gesture of panache, the artist treated us to trompe-l'oeil imitations of water drops—an odd mannerism to include in a monochrome print.

This style of flower portrayal was not common in lithography. Its use in this particular example provides an opportunity to see the medium used especially well. Although achieved by quite different means, it makes a striking comparison with the stipple engravings prepared by Ruotte and Gouy (items #24, 25). In general the application of lithography to botanical illustration followed a quite different stylistic course. Lines and tones were reduced to fairly simple formulae to allow for the hand coloring that assumed such importance in botanical illustrations.

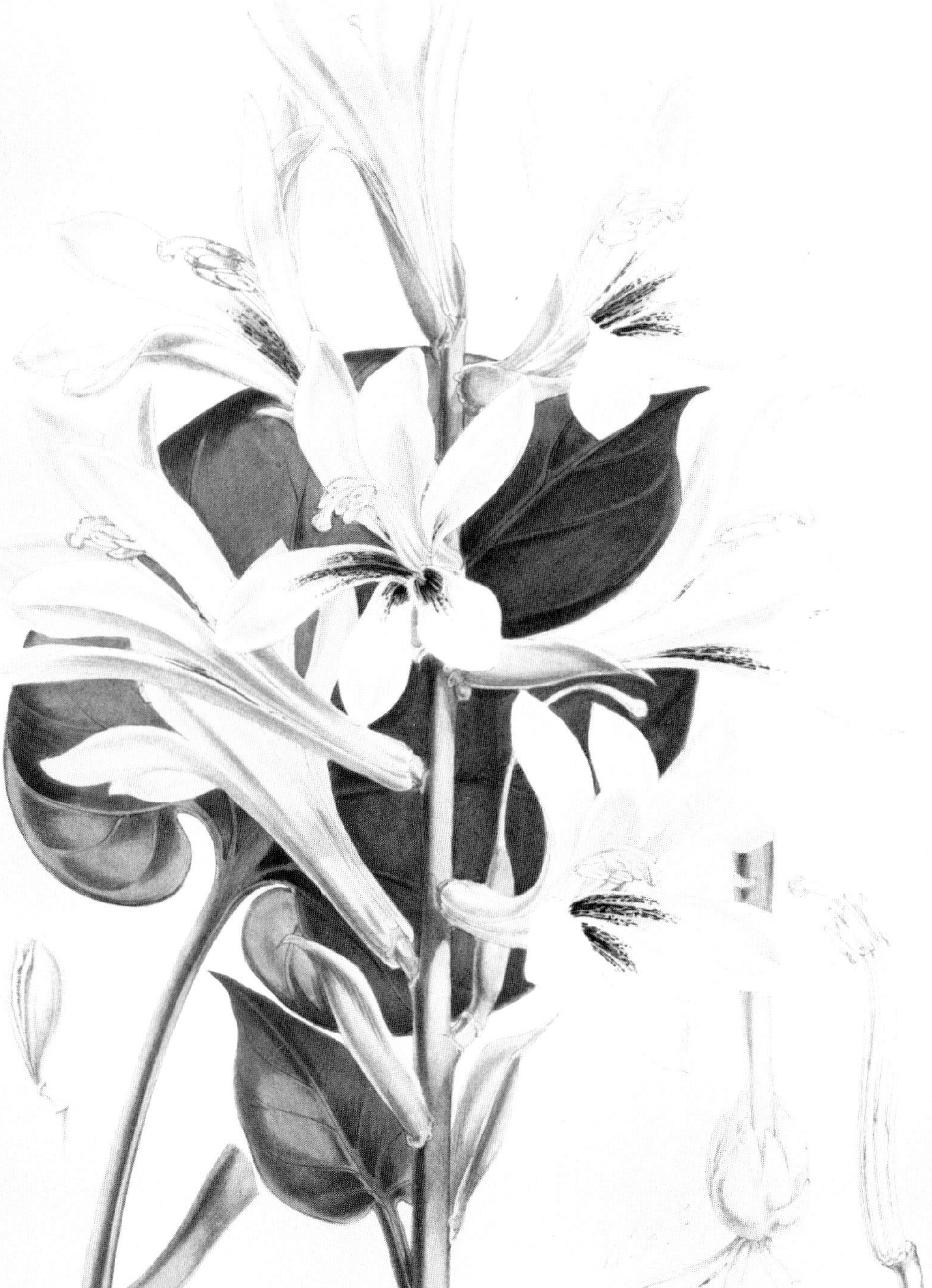

Lilium Cordifolium

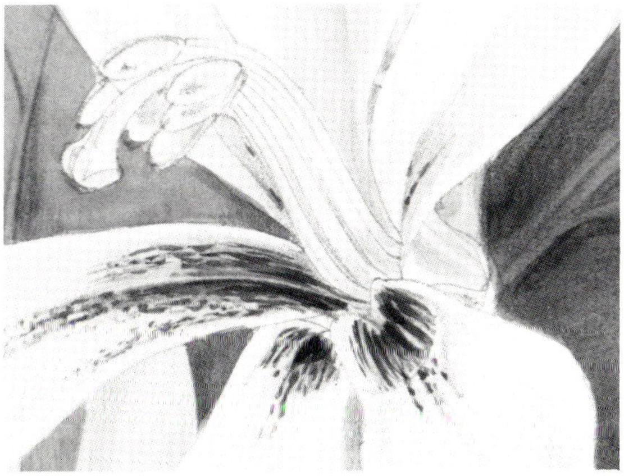

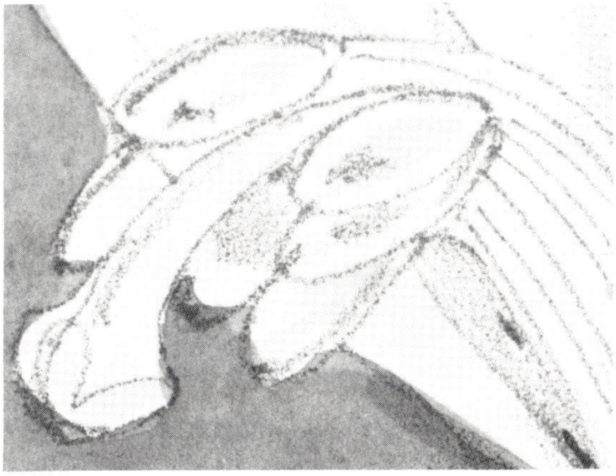

53. Lilium Cordifolium. Lithograph, chalk style, outline and chalk shading, drawn on stone by Walter Hood Fitch (English 1817-1892) after his own original. Forming a full-page illustration printed at London in 1878, lithographic printer unknown. Printed in black ink (image area 49.5 x 33.8 cm.) on wove paper (top side, sheet size 56 x 38.5 cm.). Hand-finished in transparent and opaque watercolor. Ex: H. J. Elwes. *A Monograph of the Genus Lilium.* [London, Taylor & Francis], [1877-]1880. Plate 1. Nissen BBI #594.

Chalk-style lithography achieved immense popularity as a medium for book illustration in the second quarter of the 19th century. Natural-history books accounted for a significant portion of that and naturalists remained loyal to the medium until the early 1900s, long after most other literatures had changed to modern processes. Much of the credit for the conventional style of botanical lithography, and indeed a significant portion of its sum total, was due to one man, W. H. Fitch. His output amounted to nearly 10,000 published drawings over some 46 years of activity.

Much of Fitch's work was done in a small format, especially his drawings for the Botanical Magazine, *but when called upon he produced some outstanding plant portraits for stately folios. This example, from his later years, is one of four dozen done for Elwes' treatise on lilies.*

It is hardly surprising that Fitch's immense productivity should have obliged him to develop a more sparing style than that used by Tournier for his bouquet. Fitch's approach was quite different. He sketched plants with swift, unerring outlines and then shaded economically with chalk. Much of his work was prepared for hand coloring and contained broad areas of white space. W. B. Hemsley, who saw him at work on these lily plates, remarked that he worked on stone "without hesitation, and with a rapidity and dexterity that was simply marvellous." W. J. Hooker, who first discovered and nurtured Fitch's talent, praised his "unrivalled skill in seizing the natural character of a plant." His uncomplicated style was almost certainly an important factor in achieving this.

The lithographic chalk line could not rival the minutely fine one of the etching needle or the burin in depicting small structures. In order to satisfy the botanist's need for precise information on critical characters, Fitch generally provided boldly drawn enlargements of the floral parts.

Working at one of the great centers of botanical research and with a record of 40 years of contributions to the universally respected Botanical Magazine, *Fitch was strongly influential in establishing, in Britain at least, a standard style for botanical lithographs.*

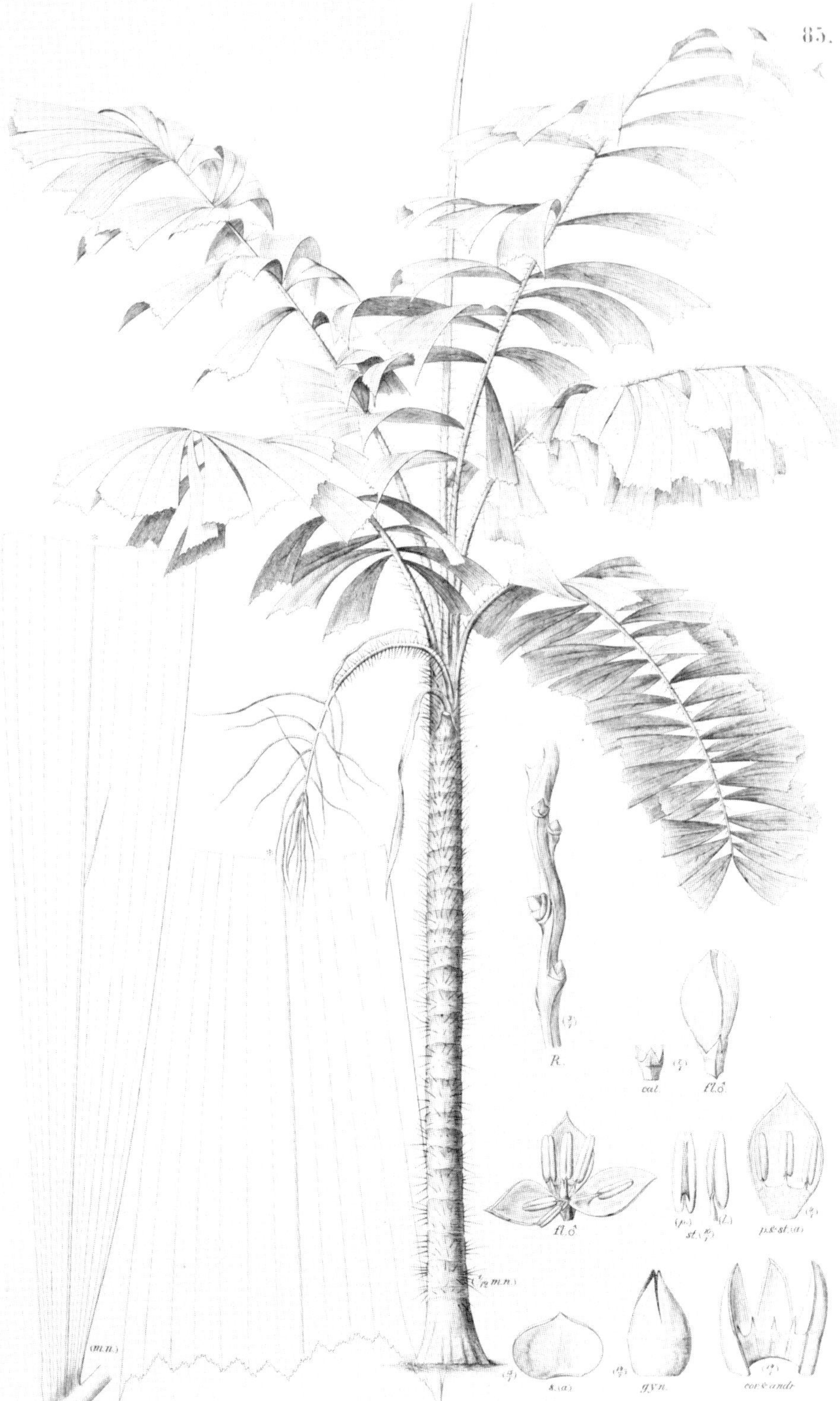

MARTINEZIA caryotifolia.

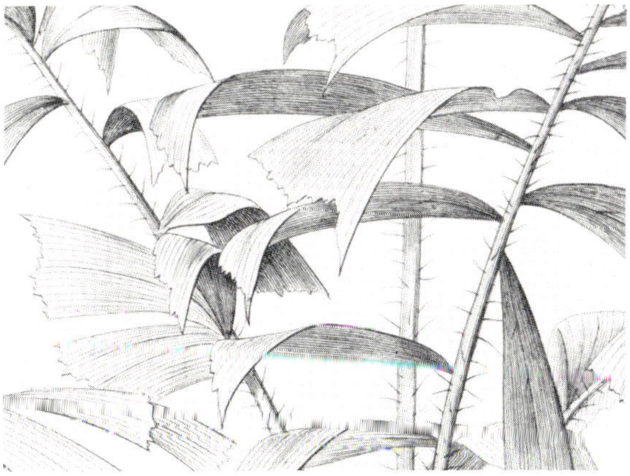

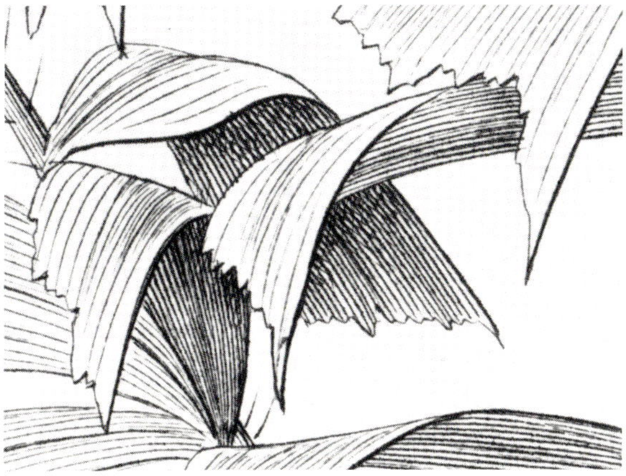

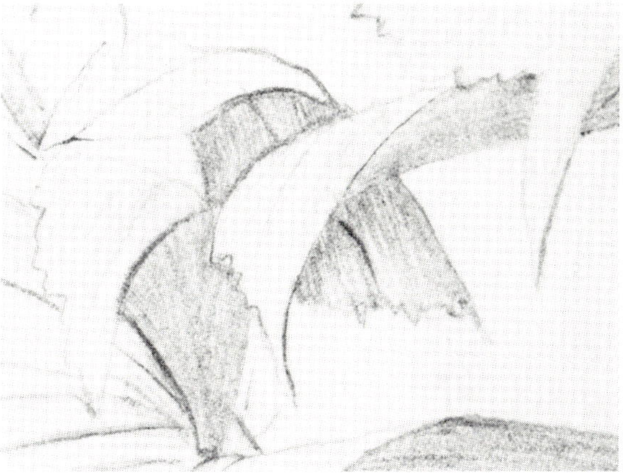

54. MARTINEZIA caryotifolia. Lithograph, line drawn, with some cross-hatched shading, drawn on stone, lithographer unknown, after an original by Jenny Bartling (?German fl. ca.1891-1930). Forming a full-page illustration printed at Munich ca.1891, lithographic printer unknown. Printed in black ink (image area 39.5 x 24 cm.) on wove paper (wire side, sheet size 49.2 x 31.5 cm.). Ex: K. F. P. von Martius, ed. *Flora Brasiliensis. Cyclanthaceae et Palmae.* By A. Drude. Munich, etc., Martius. Volume 3, part 2, 1881. Plate 85. Nissen BBI #2248. Shown together with the original pencil drawing (image area 39.5 x 24 cm.) on wove paper (sheet size 44 x 27.8 cm.).

Lithography allowed a wide range of treatments. Lithographic stone could be engraved with needle-fine linear work, drawn or stippled upon with pen-and-tusche (lithographer's drawing ink), or painted with washes of tusche. It could accept designs transferred from engraved plates, from drawings made on textured paper, or from mechanically prepared shading mediums. Lithographic chalk was available in various degrees of hardness; copal chalk, the hardest, was designed for very fine line drawing. Furthermore, grained zinc could be substituted for stone with almost indistinguishable results.

The chalk style favored by Fitch and his English followers was not universally adopted elsewhere. The 3811 plates prepared between 1840 and 1906 for the monumental Flora Brasiliensis *included this lithographic portrait of a palm species, and we are fortunate in having the artist's original drawing for comparison. The difference in style from Fitch's lily is immediately apparent. Bartling's pale and rather weak pencil drawing was quite transformed by the German lithographic draughtsman, whose copy on stone has a formal, sharply linear quality more usual in older engraved work. All the modelling was achieved by line shading without a trace of chalk work; some of the lighter shading has hair's-breadth fineness. In fact, Bartling owed quite a lot to this nameless lithographer who added vigor and definition to the whole image without alterating any botanical details.*

The scale of Bartling's habit study of this large plant allowed little opportunity for the depiction of critical taxonomic detail. She was obliged to crowd her illustration with a dozen additional drawings of magnified structures, in which her linear manner permitted precise delineation of all details. The final plate has a much more diagrammatic quality than earlier exhibits, but still it ably satisfied the needs of the scientific readers it served.

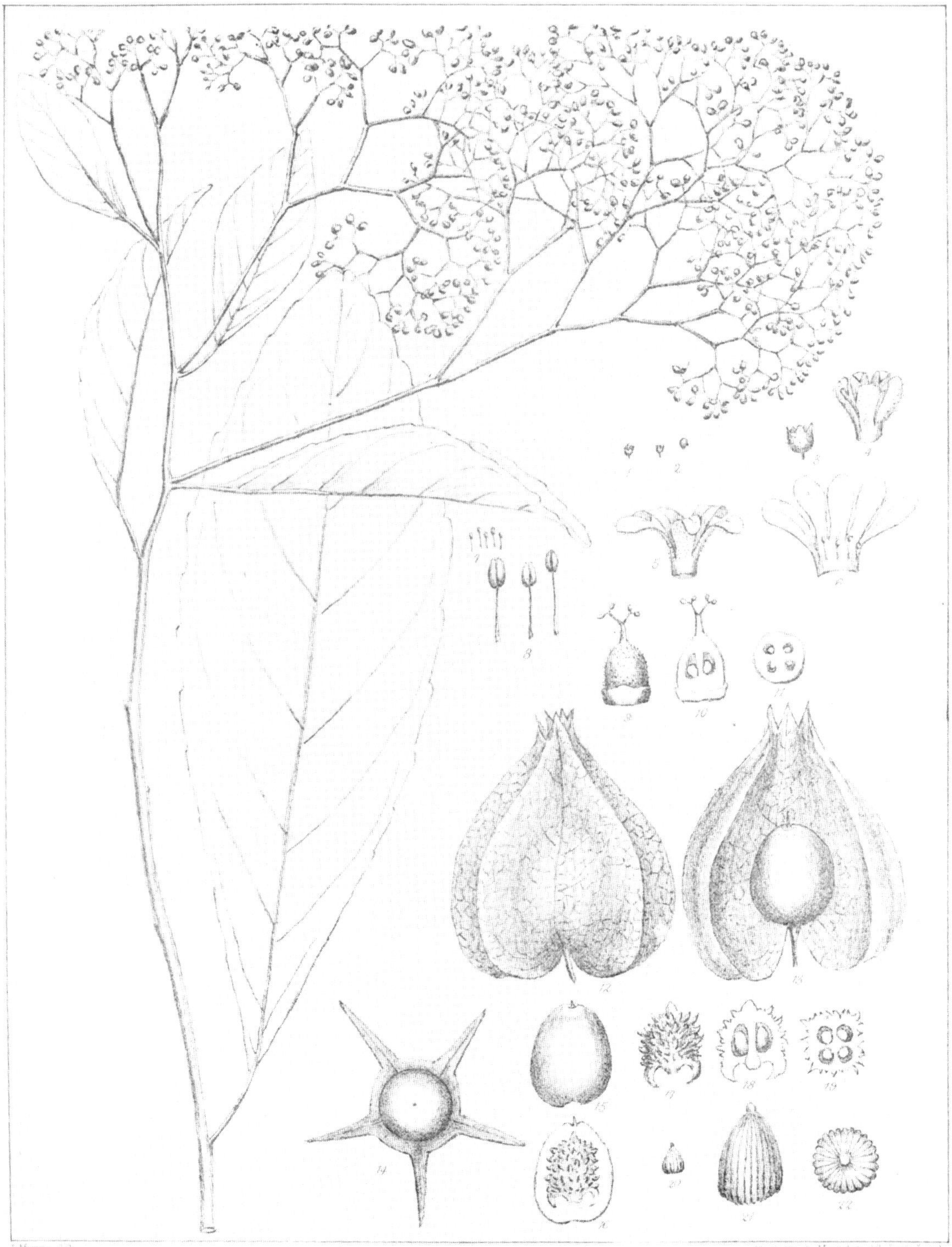

J. Miers del.

Mintern & Macdonald Lith. London

AUXEMMA GARDNERIANA.

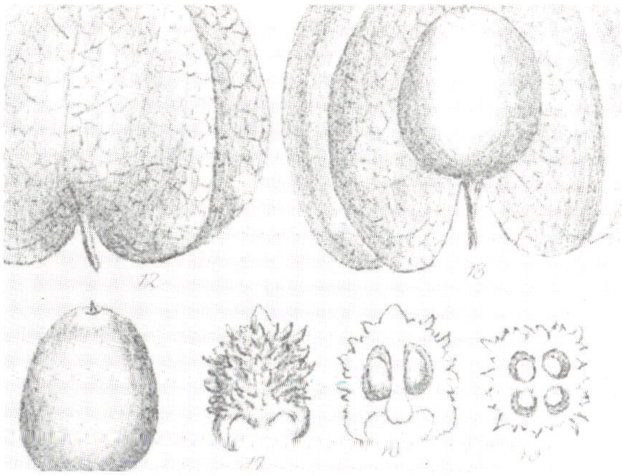

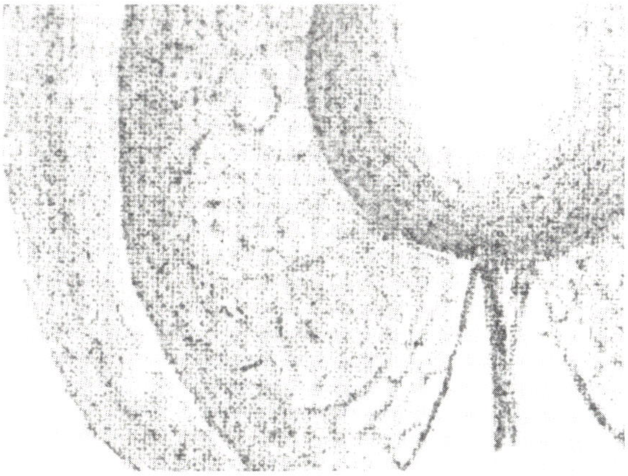

55. AUXEMMA GARDNERIANA. Lithograph, line drawn, transferred from a drawing made on textured paper by John Miers (English 1789-1879) after his own original. Forming a full-page illustration printed at London in 1875 by Maclure & Macdonald. Printed in brown ink (image area 26.8 x 20 cm.) on machine-made wove paper (matte finish, sheet size 30 x 23.5 cm.). Ex: Linnean Society of London. *Transactions... series 2. Botany.* London. Volume 1, 1875. Plate 5. Lent anonymously.

This plate from the Linnean Society's Transactions *is an early example of the unashamedly diagrammatic style of botanical illustration that later gained a firm foothold in research literature. Crowded into a corner of the plate, the sketch of the whole plant follows the geometry of the border, apparently depicts a herbarium specimen rather than attempting to show growth habit, and takes second place to the 22 drawings of enlarged morphological details. Being wholly the work of the botanist who wrote the accompanying article, the plate has unquestionable scientific authority.*

W. H. Fitch almost always drew his own illustrations on stone. Indeed, lithography had the popular reputation of permitting the artist this autographic involvement. In practice though, it could be a tricky medium, since drawings had to be made in reverse, and relatively few artists succeeded in avoiding several other troublesome pitfalls so graphically illustrated in C. J. Hullmandel's Art of Drawing on Stone *(1824). Most 19th-century lithographic illustrations were signed both by the original artist and by a professional lithographic artist employed by the printer to copy original drawings onto stone. Later in the century the numbers of these trade artists increased and their status decreased, to the point that they generally ceased receiving credit in the signature line. The preceding print (item #54) is unsigned and probably was drawn on stone by just such a professional draughtsman.*

John Miers, an engineer by profession, was a gifted amateur botanist and a competent draughtsman. He drew this illustration on mechanically embossed transfer paper with lithographic chalk. The peaks of the embossed grain acted like the "tooth" of a lithographic stone, giving a granular quality to the chalk line. The drawing was placed face-downward on a stone and passed through the press. The greasy drawing transferred a mirror image of itself that was fixed, inked and printed like a normal lithograph. Thus Miers' every line was faithfully reproduced without his having to master the technicalities of drawing directly onto stone.

Transfer paper had been introduced long before, but this particular type was developed in the early 1870s by Maclure & Macdonald, who printed the illustration. By printing the plate in brown ink they succeeded in deemphasizing the mechanical grain, which otherwise would have lessened the acceptability of this novel technique.

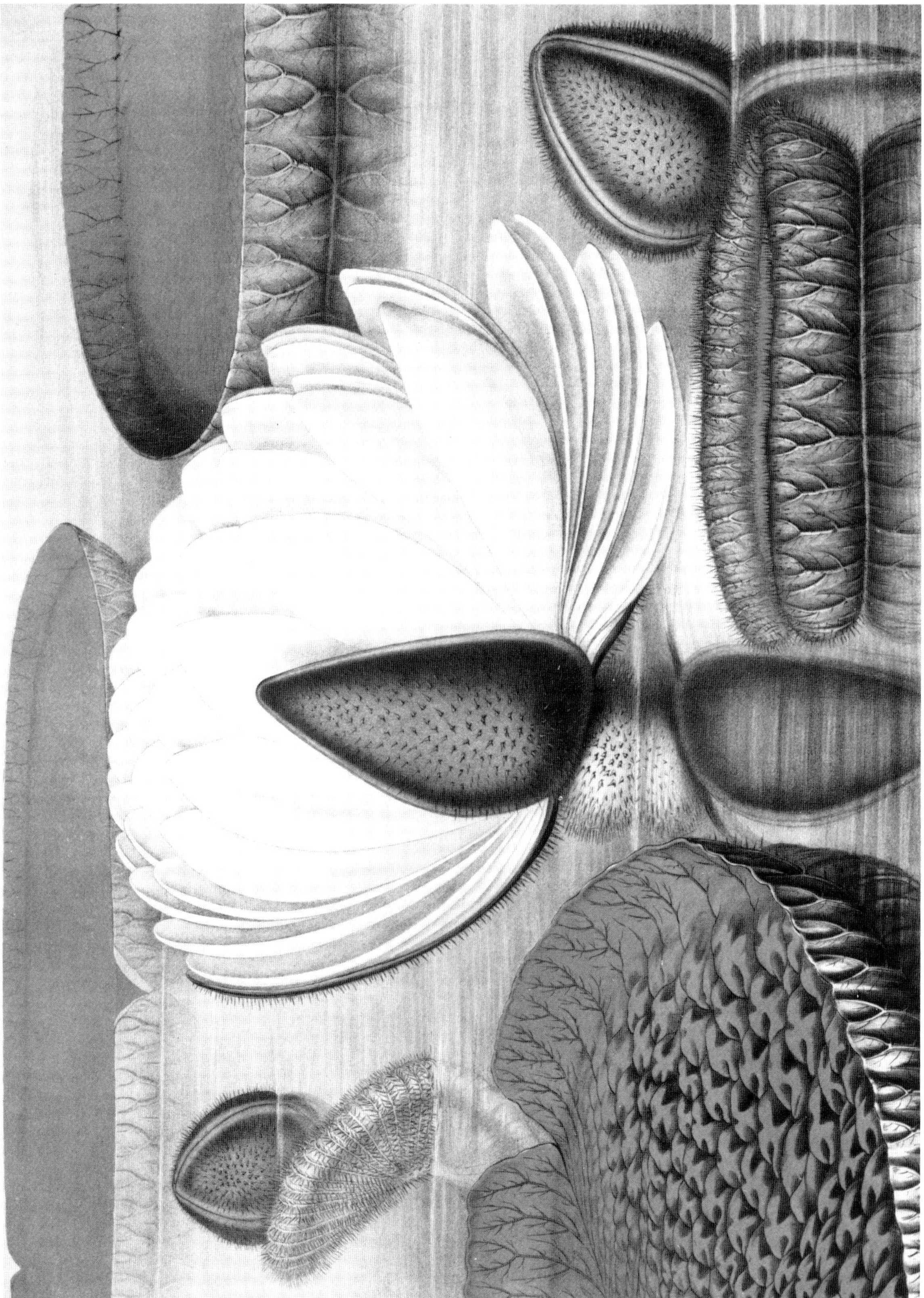

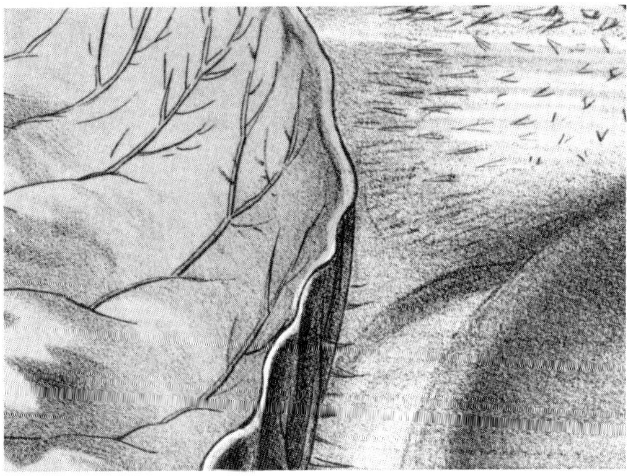

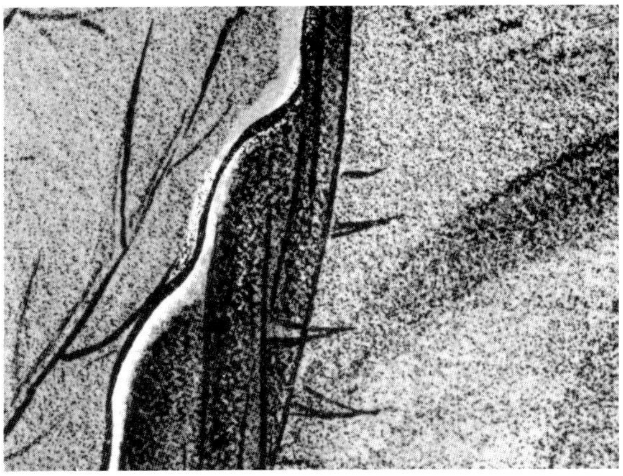

56. [Victoria amazonica]. Chromolithograph, chalk style, with a "key" stone and superimposed tints in tusche and chalk on four color-separated stones, drawn on stone by William Sharp (English/American ca.1802-post1862) after his own original. Forming a full-page illustration printed at Dorchester, Mass. in 1854 by Sharp & Son. Color-printed in black, blue, russet, straw and grey-green inks (image area 38.5 x 53.5 cm.) on wove paper (top side, sheet size 56 x 73.8 cm.). Ex: J. F. Allen. *Victoria regia*. Boston, the Author, 1854. Plate [2]. Nissen BBI #16.

Colored pencil sketches were much in vogue during the 19th century, and the close similarity between chalk-style lithography and pencil drawing invited the addition of hand coloring. Fitch's lily is typical of thousands of botanical lithographs colored that way. Transparent coloring permitted the chalk work to show through and preserved the modelling of contours.

Lithographs could be printed in any color, not just black. In the mid-1830s some early attempts were made at coloring lithographed drawings by over-printing tints from separately drawn stones. In its simplest form, this new "chromolithography" merely added flat colors over certain passages of the drawing in the manner of simple watercolor washes. Separate stones were used for each color, and the tints were prepared using chalk or tusche and could include white highlights for added effect. Secondary colors could be produced by overprinting two separate tint stones.

The chief difficulties with chromolithogrpahy lay in drawing each patch of tint so that it matched exactly the area to be colored, and in achieving perfect register with the "key" image. At that time all paper had to be damped before printing, which almost always caused its surface to expand in relation to the degree of moistness. Keeping the dimensions constant throughout the series of separate printings required for chromolithography posed real problems for the printer. Failure to do so caused misregistration of colors.

This example is the work of America's first chromolithographic printer and is one of six plates produced for a monograph on these giant plants. It displays a classic chromo-tinted style. The basic drawing in black chalk is sufficiently complete in line and shading to stand on its own. The carefully printed tints color it but add scarcely any modelling. Much of the background is grey-green tusche, heightened with a blue chalk tint for the water. A straw tint was added over parts of the print, including all but the most distant of the leaves. A solid blue tint also added over the leaves combines with the straw to create green. On the distant leaf the blue tint overlies grey-green to create a different shade of green, and so on, the four colors having been used with creative effect throughout. The result looks much like a watercolored pencil drawing except for the flatness of some of the green. Three corners of the image area have pin holes, evidence of the means by which the printer obtained such good registration of the colors.

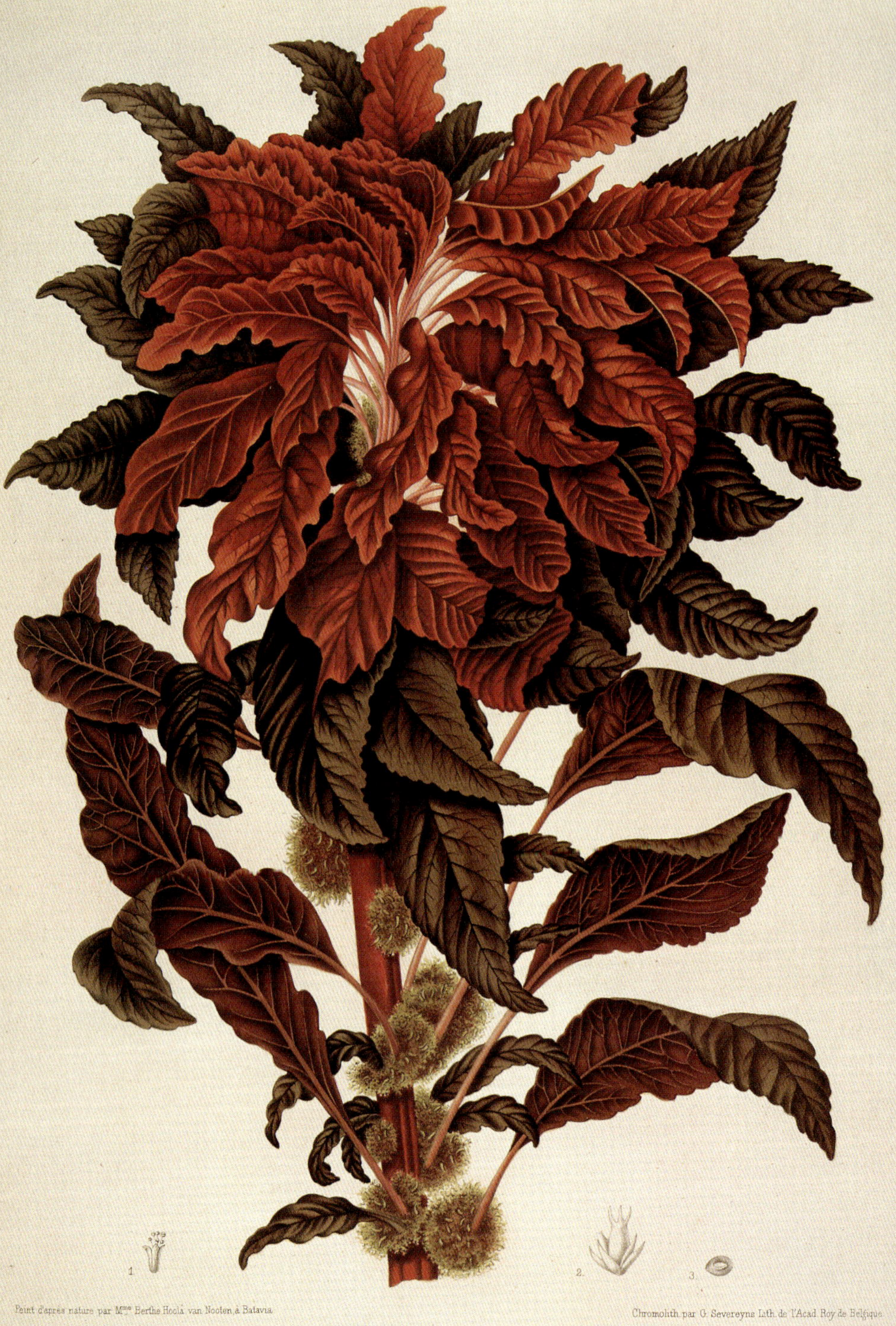

Peint d'après nature par Mme Berthe Hoola van Nooten, à Batavia

Chromolith. par G. Severeyns Lith. de l'Acad. Roy. de Belgique

AMARANTUS TRICOLOR L. BAYEM MEERA.

Emile Tarlier éditeur, à Bruxelles

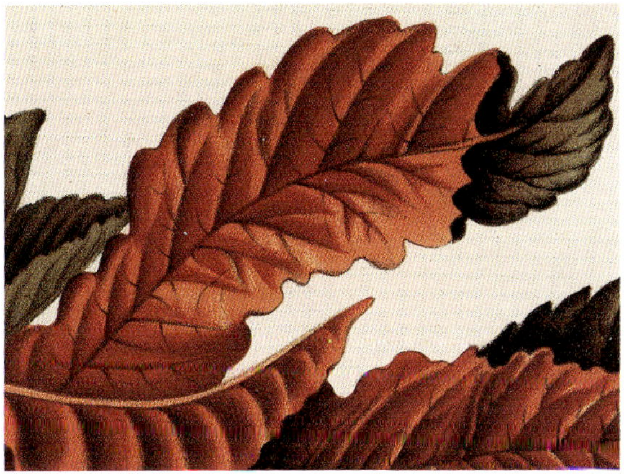

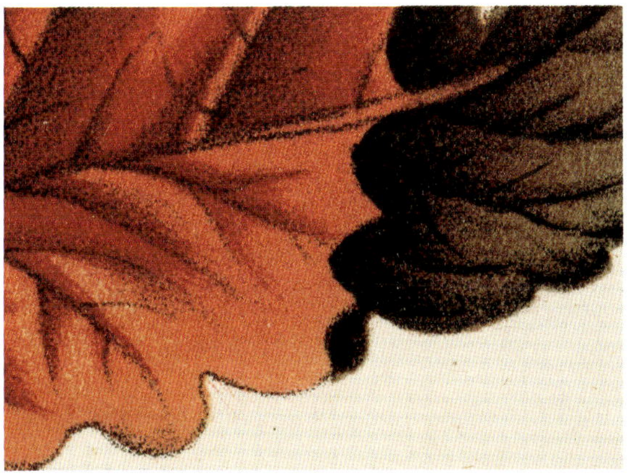

57. AMARANTUS TRICOLOR L. BAYEM MEERA. Chromolithograph in chalk and tusche, on four color-separated stones, lithographic draughtsman unknown, after a painting by Bertha Hoola van Nooten (Dutch fl. ca.1862-1873). Forming a full-page illustration printed at Brussels ca.1863 by G. Severeyns. Color-printed in red, violet, brownish-grey and black inks, the image overprinted with a clear varnish (image area 45.3 x 29.5 cm.) on machine-made wove paper (calendered finish, page size 57.8 x 43 cm.). Local details hand-finished with transparent water-color. In: B. Hoola van Nooten. *Fleurs, Fruits et Feuillages...de l'Ile de Java.* Bruxelles, Muquardt, 1863[-1864]. Plate [?]. Nissen BBI #931.

All techniques of lithography improved greatly as the 19th century progressed, and several new ones were developed. One of the most significant trends was the replacement of the older chromo-tinted style with full-color chromolithography.

Printers' artists steadily acquired greater skill in drawing clever separations to create elaborately colored images with only a few primary tints. They were able to produce chromolithographic facsimiles of paintings without using "key" drawings; the whole would be conceived entirely in colors. The discovery of aniline and other synthetic dyes much enlarged the available range of hues and intensities of pigments and provided the color artist with even greater possibilities. Meanwhile, improvements in the precision of machinery ensured consistent accuracy in registering multiple printings.

Belgium and Holland, both countries with flourishing horticultural industries, produced some magnificent chromolithographed flower and fruit illustrations. Then as now, horticultural sales were enhanced by convincing colored pictures and printers were paid high prices for producing them. This striking chromolithograph is one of the series of florid plates of Javanese plants painted by Mrs. van Nooten, about whom virtually nothing is known. G. Severeyns, who printed them, was probably the leading European color printer serving the horticultural trade.

Severeyns' lithographic artist created this picture using red, violet, brownish-grey and black inks, mostly in tusche but with some local touches of chalk. He analyzed the constituent colors and their values in the original painting and prepared a separate, detailed lithographic drawing for each. Apart from providing self and combining color to help create the full range of hues, each separation had to contribute some of the modelling necessary to give the completed image its contours and perspective. The uncredited lithographer succeeded remarkably well with only four hand-drawn separations.

The color-printed image was overprinted with clear varnish. It enhanced the depth of color overall and added finish to an extraordinarily vivid and imposing plant portrait.

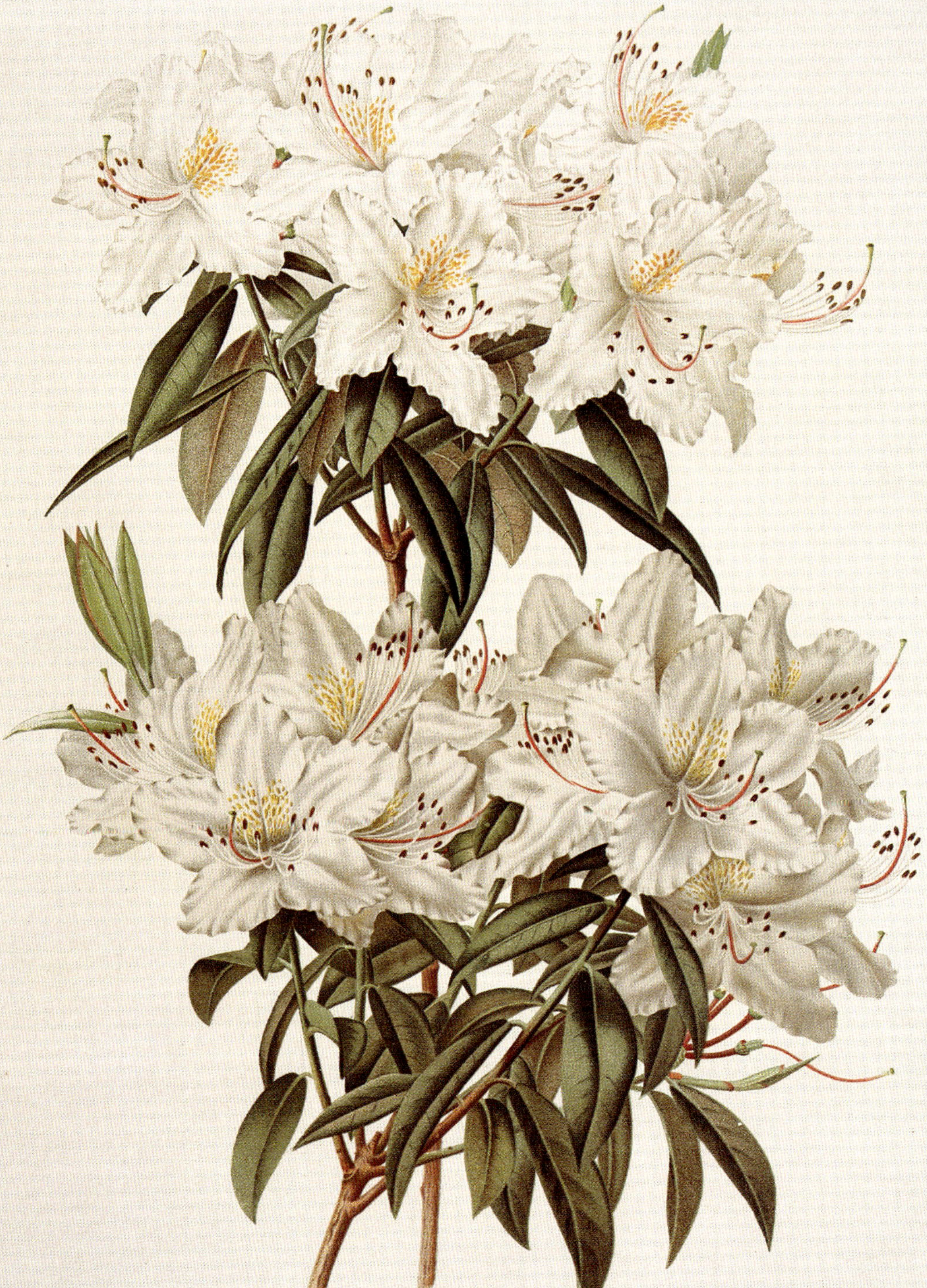

RHODODENDRON AUGUSTINII AND ITS WHITE FORM

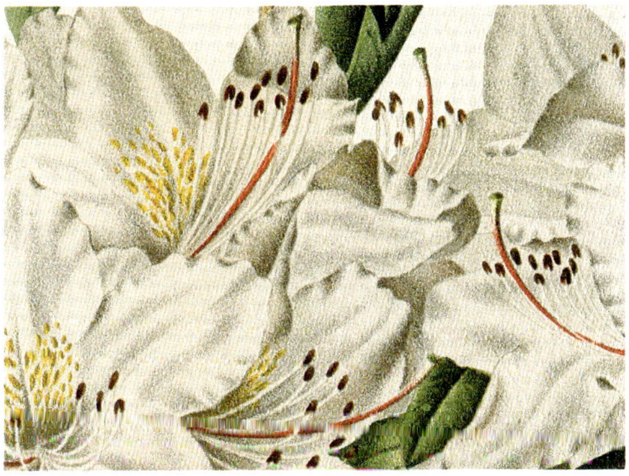

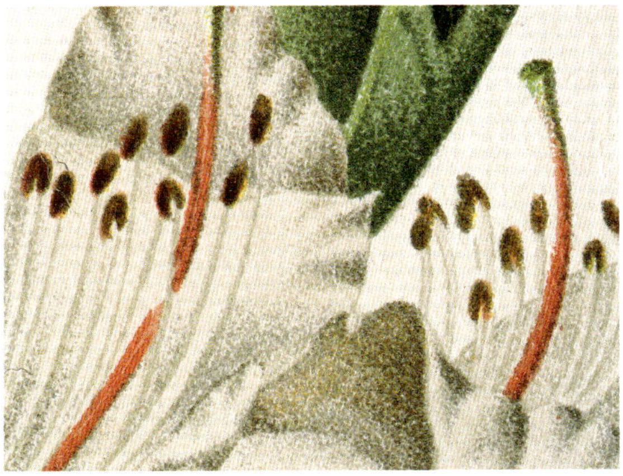

58. RHODODENDRON AUGUSTINII AND ITS WHITE FORM. Chromolithograph, stippled manner, on six color-separated stones, from "a drawing kindly supplied by M. Maurice de Vilmorin," lithographic draughtsman and artist unknown. Forming a full-page illustration, printed in 1905, printer unknown. Color-printed in yellow, peach, dove grey, green, red and very pale grey inks (image area 28.2 x 20 cm.) on machine-made wove paper (calendered finish, impressed texture applied after printing, page size 31.4 x 23.5 cm.). In: W. Robinson, ed. *Flora and Sylva*. London, the Author, 1903-1905. Volume 3. Facing page 162. Nissen BBI #2251.

Towards the end of the 1870s, lithographic machinery capable of relatively high-speed operation was introduced on a wide scale. Until then the best chalk lithography had been printed on hand or power assisted lithographic presses, and skilled pressmanship was the factor that ensured fine results. But the arrival of mass production in response to increasingly voracious markets eclipsed individual pressmanship.

Another significant loss was that delicate chalk style could not be printed on the new machines. The pressure imparted by their rotary cylinders was inadequate to print grained stones cleanly. It became necessary to grind and polish stones to a shiny smoothness and, without any "tooth" on the stone, the crayon was deprived of any chalky character. One remedy was to transfer drawings from roughly textured or patterned transfer paper, as seen in item #55. Another was to create chalk "effects" by stippling. Very fine pen-and-tusche stipple could be used to create subtle tones and shades just as chalk-style engravers had done in the 18th century. Although most chromolithographic work was created with washes of tusche, the stippled style became widely popular towards the end of the century and great quantities of cheap and generally poor-quality work were produced. However, in sensitive hands this method could produce very fine results, as demonstrated here.

The lithographic artist used only six colors to create this extremely skillful portrayal of a white flower on white paper. The whole image is made up of minutely "grained" stippling, and apart from achieving accurate coloration the artist provided convincing modelling. Lithographic work on the new smooth stones necessitated smooth paper for clean printing. This tended to detract even further from the acceptability of much late-19th-century chromolithographic work. This picture is printed on very smooth, slightly glossy paper. In order to conceal this somewhat unpleasant surface, the completed print was run under a strongly textured roller which left an impressed texture similar to that of good drawing paper. The identity of the printer is not known. William Robinson, editor of Flora and Sylva, *which contained 64 of these chromolithographed flower portraits, tells his readers merely that he "went with flower drawings to the best color-printer in Europe." They were well served by his choice.*

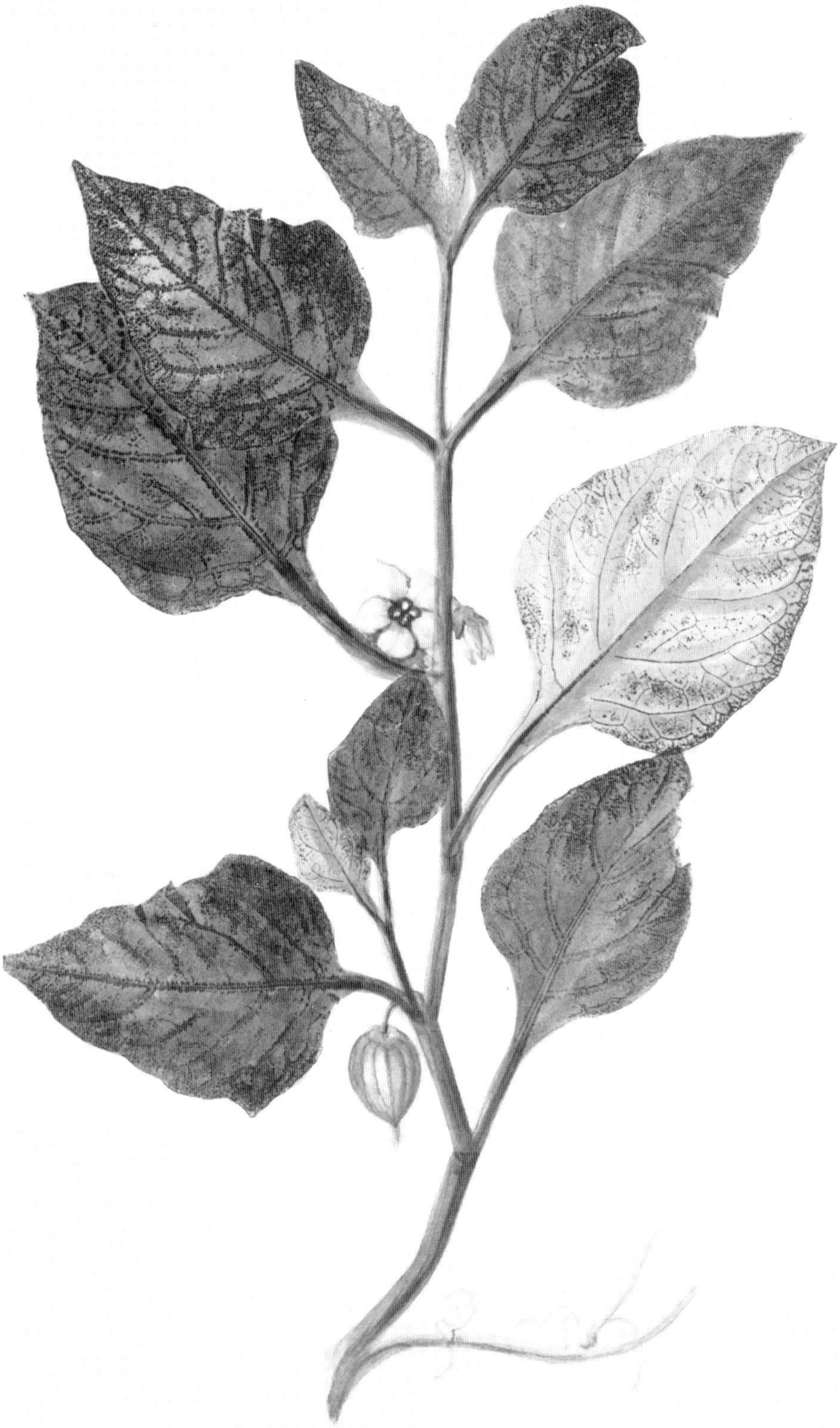

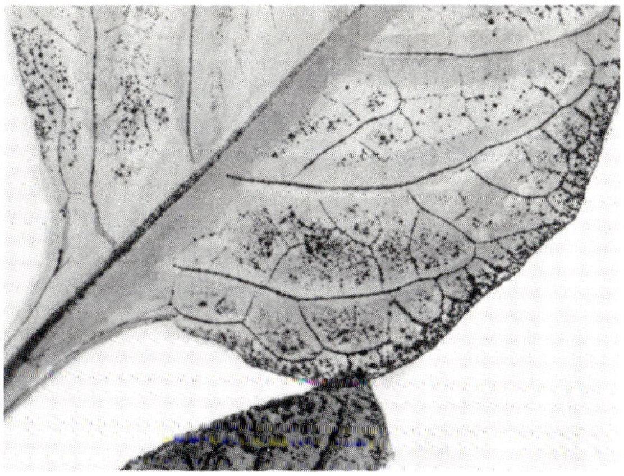

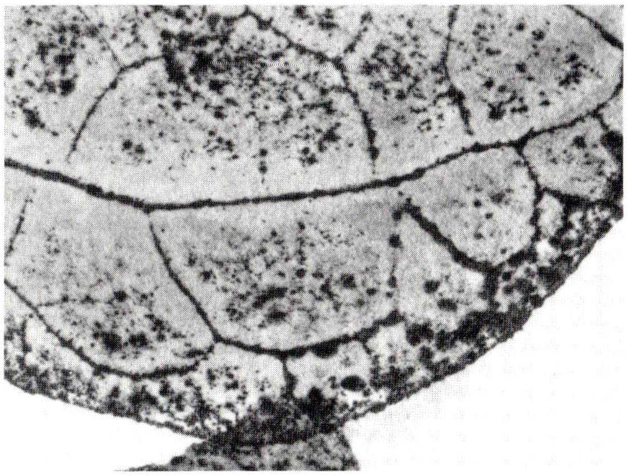

59. [Physalis]. Nature printing, printed from a flattened plant specimen coated with printer's ink. Probably printed during the 18th century by an amateur. Printed in black ink (image area 36.5 x 22.5 cm.) on wove paper (top side, folded sheet size 41.5 x 51.5 cm.). Hand-finished with transparent watercolor.

The idea of using a leaf or whole plant as the instrument for creating a separate self-image had a very ancient origin. But it was the inventive and inquisitive Leonardo da Vinci who bequeathed what is probably the oldest extant nature print on paper. His Codice Atlantico, *written between 1490 and 1519, includes a description of a nature-printing technique and a specimen image of a single leaf.*

A long, and somewhat disjointed, train of experiments followed over the next two centuries. In the late 1720s the first large-scale attempt was made at printing botanical book illustrations from the plants themselves. Leonardo had used the simple method of coating a leaf with lamp-black and oil and printing it onto the page with hand pressure. J. H. Kniphof's Botanica in Originali seu Herbarium Vivum *commenced in 1728, with nature-printed illustrations created by basically the same process. However, his prints show evidence of great care in inking the very delicate plants uniformly. The use of different specimens in separate copies of the book conforms with evidence from later experimentation that each plant yields only a small number of good impressions before becoming saturated with ink or otherwise damaged. Although the overall edition was small, producing it must have consumed a constant supply of fresh specimens.*

The print exhibited here is very similar to Kniphof's plates and was clearly made by the same process, probably in the 18th century, but its precise origin is unknown. The printer took the trouble to turn one leaf so that its underside may be compared with the upper surfaces of the others, and succeeded in highlighting its prominent venation. However, certain features were apparently unprintable by this means. The artist who hand colored the print also had to add a painted flower and fruit not printed from the specimen.

At a reasonable distance the whole effect is quite pleasant, though one is obviously aware of the flattened perspective of the leaves, which in life grew with their upper sides turned to receive the sunlight.

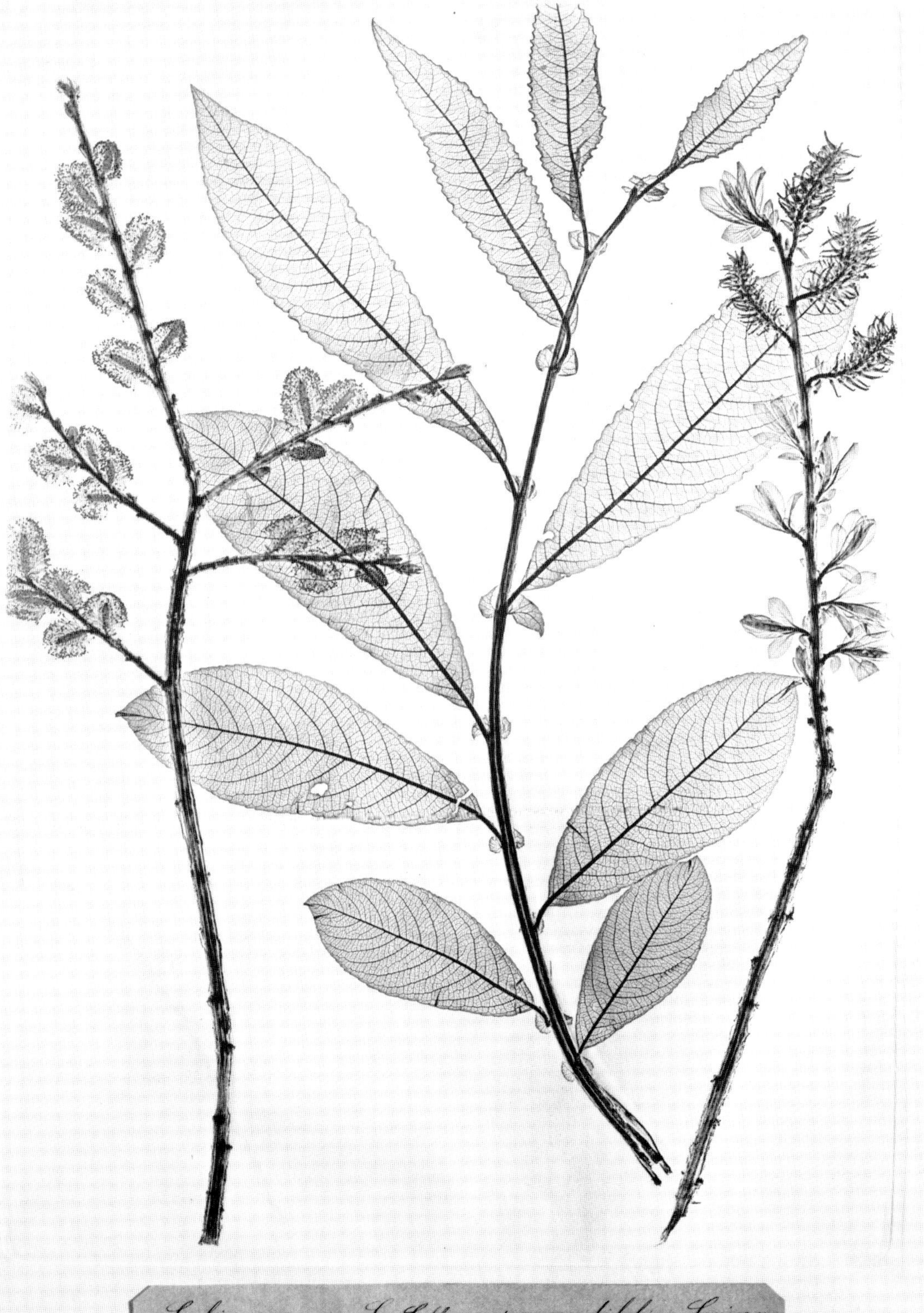

Salix (caprea L. Sahlweide) grandifolia Sering.

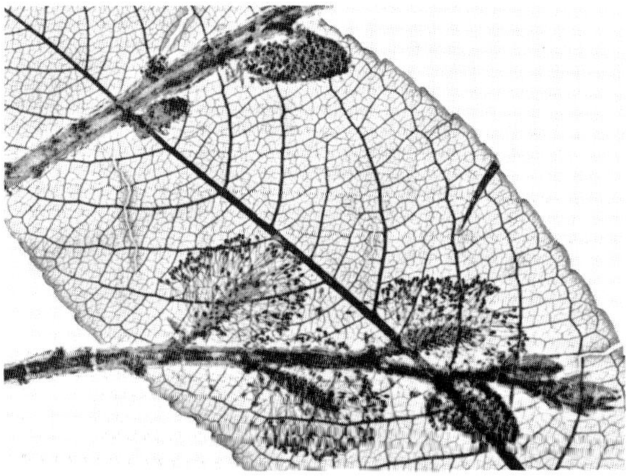

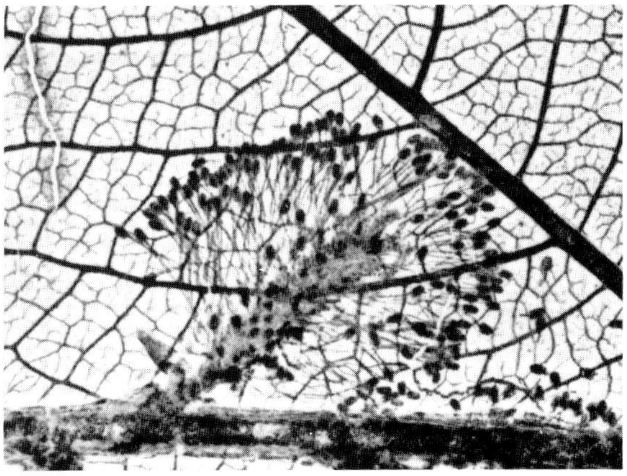

60. [Salix : Caprea L. Sahlveide : grandifolia Sering.]. Intaglio nature print, double-cast from soft-metal impression of plant (Auer's process). Printed at Vienna in 1850s? by K. K. Hof- und Staatsdruckerei. Printed in brown ink (platemark 43 x 31.5 cm.) on wove paper (top side, sheet size 57.7 x 40.7 cm.). ?Ex: C. F. von Ettingshausen and A. Pokorny. *Physiotypa Plantarum Austriacurum*. Vienna & Prague, Tempsky, [1855-] 1856. Plate [?]. Nissen BBI #614.

Nature printing has always appeared to the layman to have more potential value than it does to the botanist. One reads that it faithfully represents intricately formed plants, "the complicated forms and tender organisation of which baffle the most skillful and patient artist." But several great botanical artists have been able to depict the most startlingly complex structures with unerring accuracy in structure, modelling and perspective—the Bauer brothers, Ehret, and others spring to mind.

Thomas Moore, author of the Nature-Printed British Ferns, *was obliged to admit that "it is true that Nature-Printing has its defects as well as its advantages, for it, like the artist, can only represent a portion of the whole structure of the plant; but then its accuracy is perfect as far as it goes, and in the case of ferns, it shows just that which it is most desirable to represent for practical purposes, that is, the outline and the venation."*

The large, minutely dissected, generally flat fern frond was ideally suited to representation in this way. But three-dimensional or solid structures such as a thick stem, a bulky fruit or most kinds of flowers could only produce a squashed-looking image of themselves under pressure. Furthermore, the three-dimensional habit of a plant had to be altered so that it could be represented on a single plane. Such features as delicate floral parts, fine indumentum, fleshy tissues or even fern sori, which might be essential taxonomically, were generally unprintable in any recognisably useful form. The possibility of producing decent nature prints from such objects as a teasel head, a thistle head, a houseleek or a twig of pine was virtually unthinkable. The botanical artist could still rely on employment! Indeed, nature printing had relatively little impact on his world.

The most spectacular method of nature printing was that perfected in Vienna and London in the mid-19th century. Under huge pressure a dried flattened plant was squeezed between two plates, one of soft polished lead, the other hard as steel. The lead took a cast of the plant, even down to quite filmy petals, with, of course, the usual distortions. An electrotyped copper cast was taken from the lead, and then a reverse cast from that. The result was an intaglio plate that could be inked and printed like an engraving. If the plant was suitable and its image not too badly distorted, this process produced an attractive nature print. This example is typical of hundreds made by the Imperial Printing Office in Vienna.

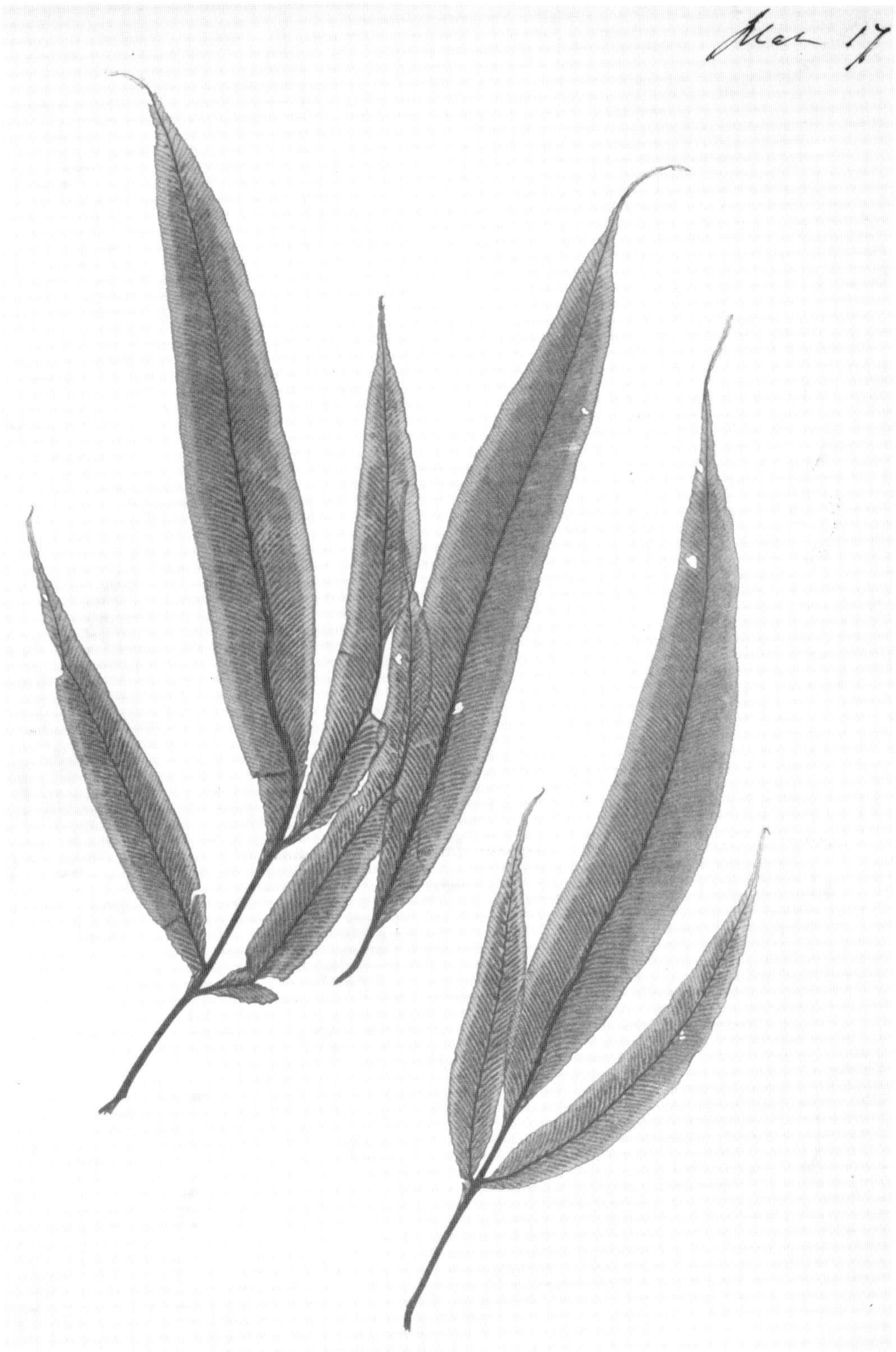
Mar 17

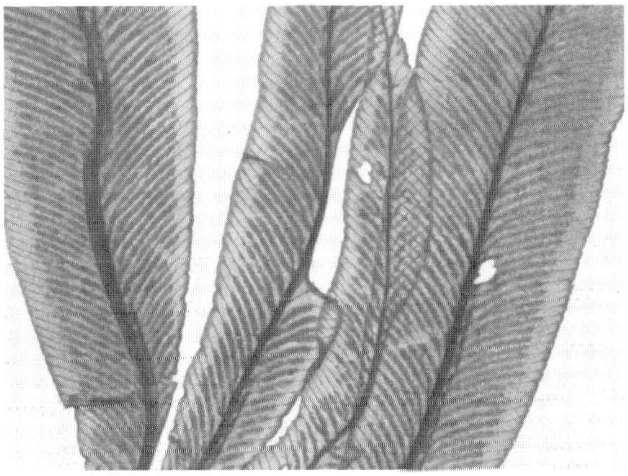

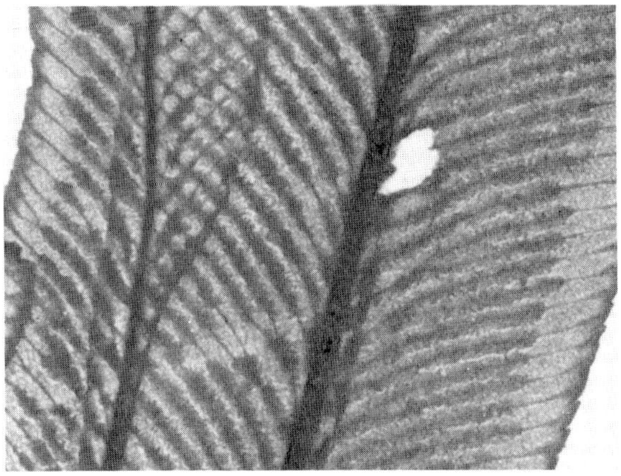

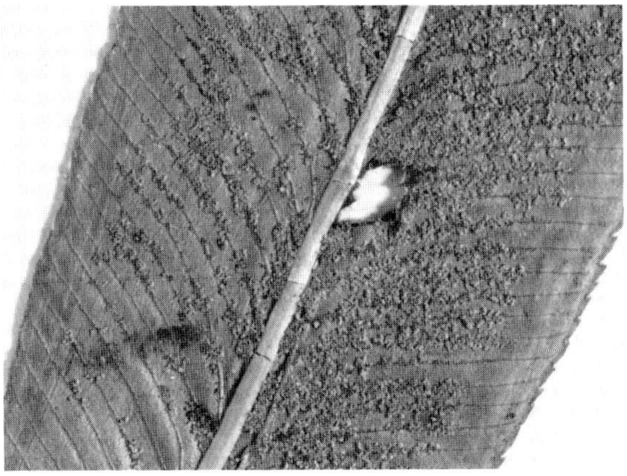

61. [Coniogramme]. Intaglio nature print, probably by double-casting from soft-metal impression of plant (Auer's process). Printed ca. mid-19th century, place of printing and printer unknown. Printed in green ink (image area 30.5 x 18.5 cm., no platemark visible) on heavy wove paper (top side, sheet size 38.5 x 28 cm.). Ex: [United States National Herbarium. Projected untitled publication. Plate 17]. Shown with the dried specimen from which the impression was taken. Indefinite loan from the Smithsonian Institution, Washington.

Impressive as intaglio nature-printed plant images were, they gave curiously misleading information. Outline and venation were represented with great accuracy, but the image created under pressure was, in reality, a picture of the relative compressibility of different plant tissues. Thus, a leaf that in life appeared thick, fleshy, opaque and indistinctly veined would be likely to give a reverse impression when dried and subjected to the great pressure of nature printing. The venation appeared as a vividly clear network, and the collapsed fleshy tissue as an ethereal film. The reason was quite simple. Fleshy tissue is mostly water and dries out to a thin cellular mesh, whereas venation comprises hard cellulose vessels that resist compression.

Surface texture is rarely discernible in nature prints. A glossy surface printed exactly like a dull waxy one. It was only rarely possible to successfully depict the presence of plant hairs. The color of the leaf and other plant tissues as they appeared in incident light could not be successfully represented. Though intaglio nature prints were produced in color by the à-la-poupée method, the results were not really convincing. Colored inks could further accentuate the ethereal quality of the print, often the reverse of what was required.

When a nature impression of a dried plant had been taken in lead, the specimen usually remained embedded in the metal. If the plate were heated, causing it to expand, the specimen usually sprang free, but brittle or tender specimens would be damaged beyond redemption. The survival of specimens from which 19th-century nature prints were taken is very rare indeed. The print and specimen shown are from a series of 23 nature prints apparently prepared for a never-published work on ferns by the United States Department of Agriculture. The plates were made by an unknown printer who used a method closely similar to the Vienna process of the 1850s and 60s.

This print exemplifies both the strength and the weakness of intaglio nature printing. When a botanist was recently asked to identify the species, he remarked that "not enough detail of the sori is available on the prints to make identification certain." This marvel of 19th-century printing is, then, little more than just that; it has little real botanical value.

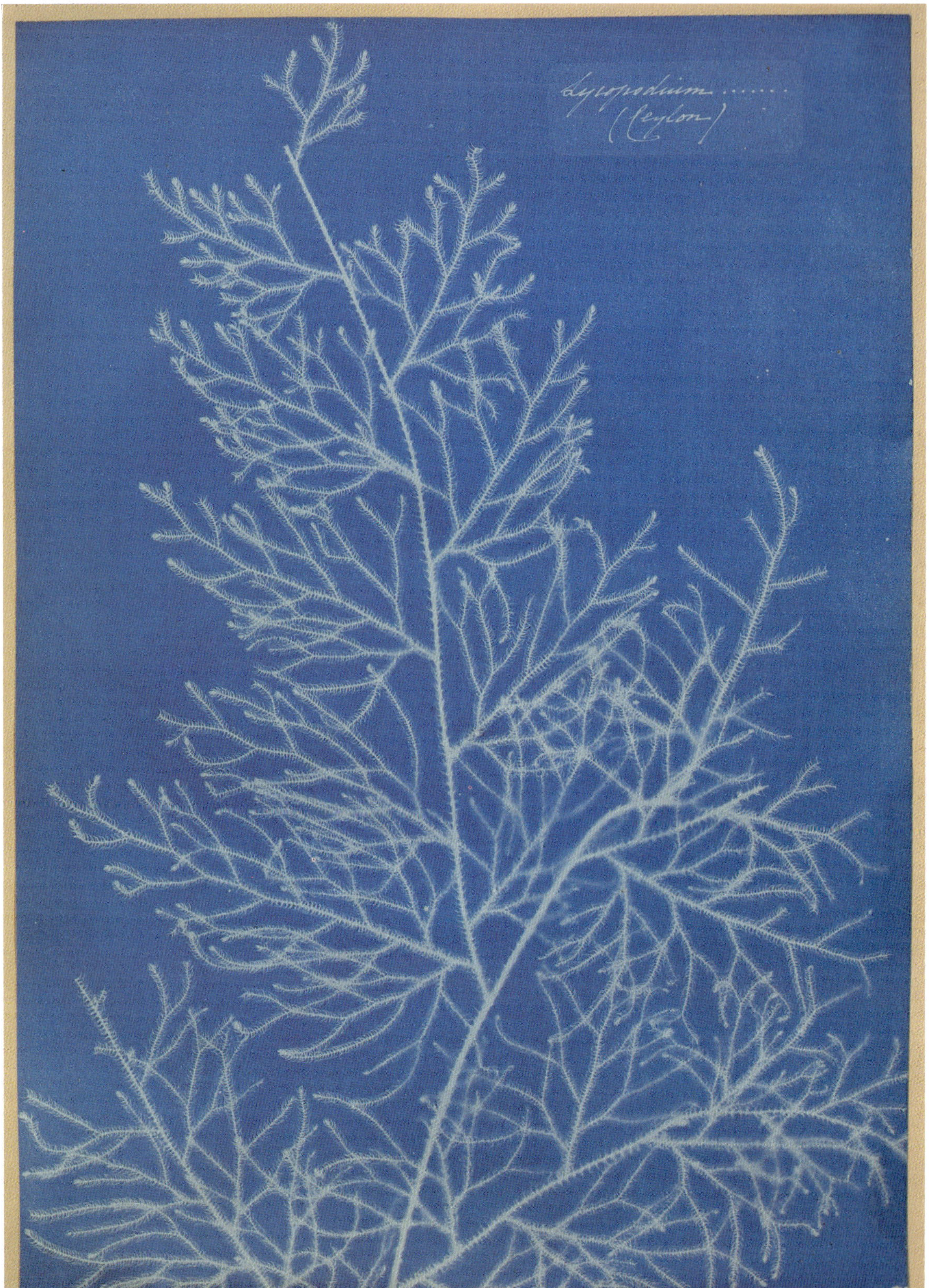
Lycopodium ......
(Ceylon)

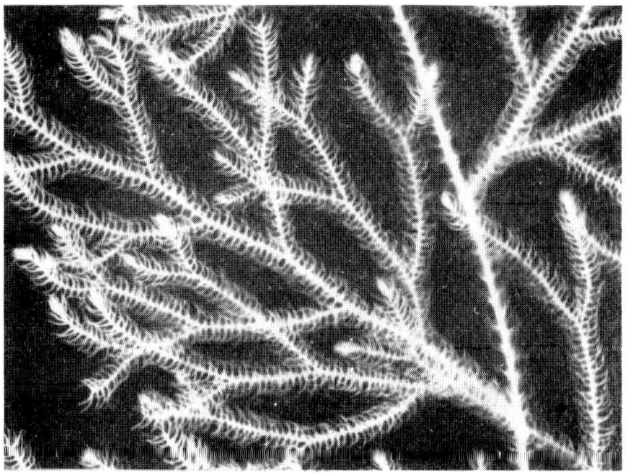

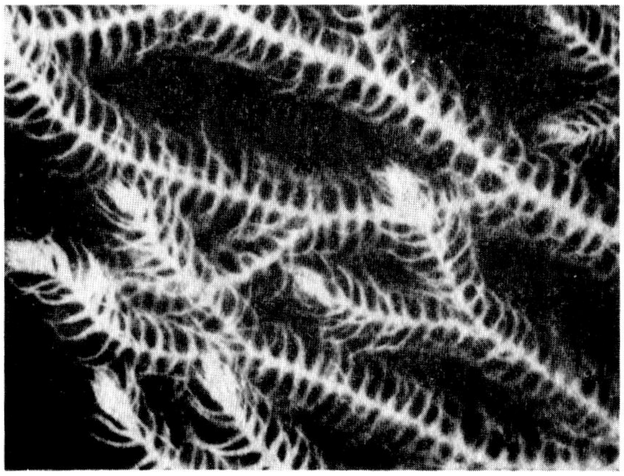

62. Lycopodium...... / (Ceylon). Cyanotype photogram of plant on sensitized paper (blueprint), made by Anna Atkins (English 1797-1871). A "negative" print on bright blue ground produced in England ca.1854. Developed on wove paper (sheet size 35 x 25 cm., laid down on buff-colored wove paper, sheet size 48 x 37 cm.). Lent by Hans P. Kraus Jr.

The arrival of photography provided the opportunity for a different form of "nature printing," one in which light was the medium. It found its fullest expression through the endeavors of Anna Atkins, an English amateur, who made photograms of plants. By this means she produced the first photographically printed and illustrated book of any kind.

A photogram is a photograph produced without a negative or camera, made by allowing a physical object to cast its shadow on the recording surface. The process that Mrs. Atkins used for her photograms was cyanotype, invented by Sir John Herschel in 1842. She laid a pressed plant on a sheet of paper impregnated with iron salts and then exposed it to light. When paper was washed in fresh water the exposed areas turned a vivid insoluble Prussian blue. Herschel's process remained in use for copying drawings and plans until recent years under the familiar name of "blueprint."

Atkins directed her industry and manipulative skills at the production of a three-volume iconography of British seaweeds. Each photogram showed the outline and apparently ethereal substance of the seaweed against a blue ground that added an appropriately aquatic ambiance. Publication commenced in 1843 and the work's more than 400 plates were to demand 11 years of sustained endeavor in collecting and preparing specimens and making separate photograms of each. Because a photogram has no negative, each copy of the book required an individual exposure for every plate. It is hardly surprising that only 10 or 11 copies of the book have been identified to date. The pioneer application of photography to this extensive iconography created images of great significance, interest and, incidentally, charm. Once her monograph on British algae was completed, Mrs. Atkins turned to producing cyanoptype prints of ferns, grasses and other land plants. The example shown is from that period.

This was the first major incursion of photography into the field of botanical illustration, and it might seem to have posed a threat to the professions of artist and printmaker. However, the photogram depicted plants in a manner quite alien to the illustrator. The outline and tonal language of the photogram were products of the relative opacities of different plant tissues. Like the intaglio nature print, this process created a somewhat oblique representation of the plant. Artists and engravers had nothing to fear from photography yet!

THE

MAGAZINE OF SCIENCE,

And School of Arts.

No. IV.] SATURDAY, APRIL 27, 1839. [Price $1\frac{1}{2}d$.

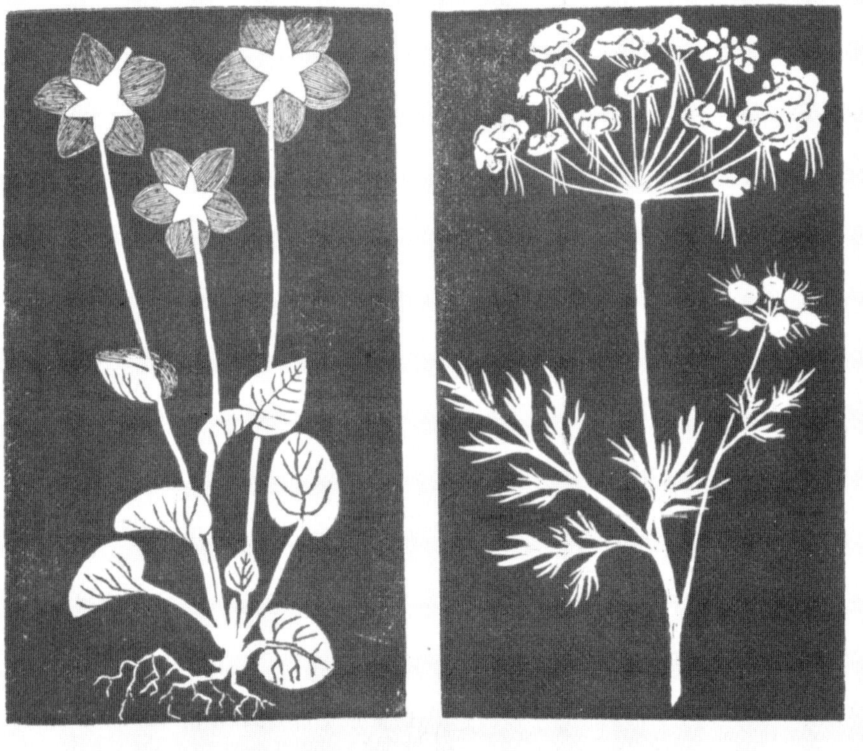

FAC-SIMILES OF PHOTOGENIC DRAWINGS.

THIRD EDITION.]

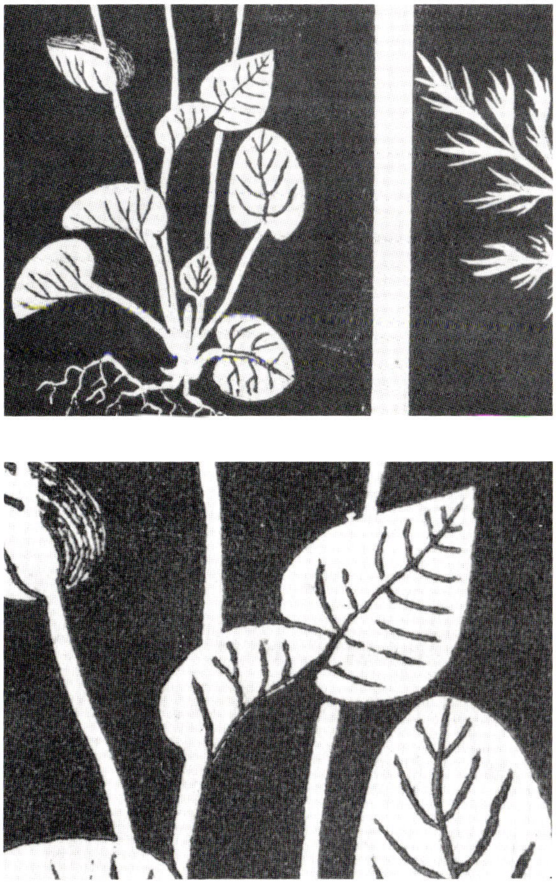

63. FACSIMILES OF PHOTOGENIC DRAWINGS. (Parnassia palustris, or Grass of Parnassus. Aethusa Cynapium, or Fool's Parsley). Wood engraving, "negative" white-line style, after a photographically produced image on the block, engraved by George William Francis (English 1800-1865). Forming an illustrated page (type area 20.8 x 12.5 cm.) printed at London in 1839. Printed in brown ink (image areas 10 x 5.8 cm. each) on machine-made wove paper (matte finish, page size 24 x 15.5 cm.). In: *The Magazine of Science and School of Arts.* London. Volume 1, number 4, 27 April 1839. Page 25. Lent anonymously.

Although Atkins' Cyanotypes of British Algae *was the first book to be illustrated and printed by photography it was not the first use of photography in connection with plant illustration. The separate discoveries of photographic processes by L. J. M. Daguerre in France and W. H. Fox Talbot in England were both published in the early part of 1839. Talbot announced his process in a public lecture at the end of January, and three months later a general review of photogenic drawing processes appeared in the recently founded* Magazine of Science.

A serious limitation of the new processes was that they could not readily provide book illustrations. Daguerreotype images were developed on silvered metal plates. Talbot's photogenic drawings were developed on paper but each copy of the book would have required a separate exposure of each print. In any case, Talbot's process was still experimental and in the inventor's possession. Many of his first demonstration prints were photograms of plants in which the images appeared in white on a brownish ground.

News of such an exciting invention needed illustration to convey some notion of its appearance. George Francis, botanist, printmaker and apparently a handy craftsman, quickly produced a solution. He photosensitized the surfaces of boxwood blocks, laid small flattened plants on them, exposed them to light and created photograms on wood. The engraver then simply cut out the images, though, as the editor remarked, "in the flowers he has failed to express the delicacy and beauty of the drawings [sic]."

These relief blocks could, of course, have been run on the printing machine alongside the text. However, obviously it was decided to enhance their similarity to Talbot's prints by printing them separately in a brown ink. To our educated eyes the results are simple and uninspiring, scarcely suggestive of photography. But, even so, their priority in photographic botanical illustration gives them special interest here.

Unfortunately, this early combination of photography and printing press was not to be repeated for many years. In 1841 Talbot took out the first of several patents for photography, by which he sought to block its use or development by others. As late as 1854, Jabez Hogg published The Microscope, *a large treatise with scores of biological illustrations, and was forced to lament that plans to use photography in preparing his wood-engraved illustrations had been thwarted by the existence of Talbot's patents. At that time Talbot had proved himself ready to go to law to protect them.*

PLATE VI.

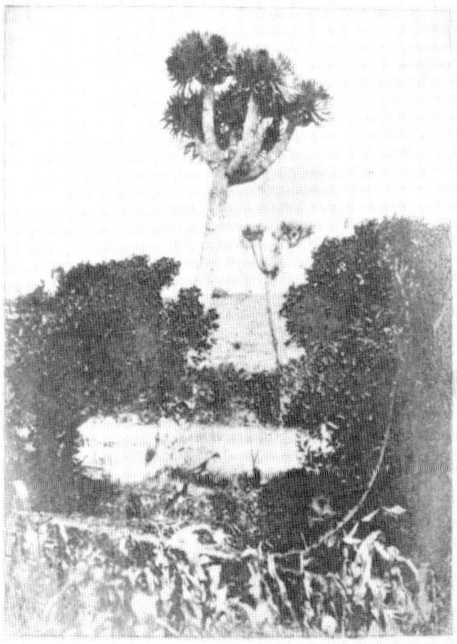

Young Dragon trees, near Orotava, Teneriffe.

Photoglyphic plate by W. H. Fox Talbot Esq.

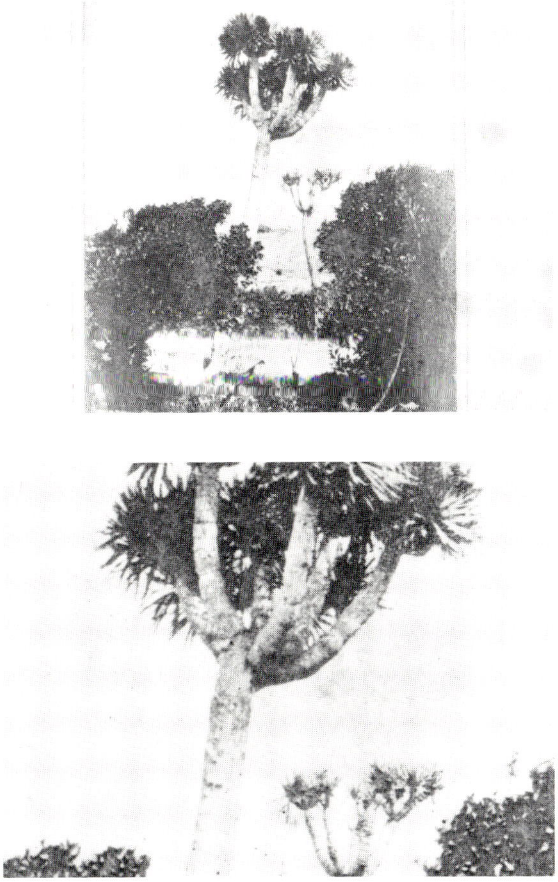

64. Young Dragon trees, near Orotava, Teneriffe. Photoengraving, etched halftone plate (Talbot photoglyph process), produced by William Henry Fox Talbot (English 1800-1877) after a photograph from life by C. P. Piazzi Smyth (English 1819-1900). Forming a full-page illustration printed at Edinburgh in 1860 by William Banks. Printed in black ink (platemark ca.7.5 x ca.5.5 cm.) on machine-made paper (page size 21.5 x 13.5 cm.). In: Botanical Society of Edinburgh. *Transactions*... Edinburgh. Volume 6, 1860. Plate 6. Lent by the Library of the Academy of Natural Sciences of Philadelphia.

Talbot first used his calotype process for book illustrations in 1844, soon after Mrs. Atkins commenced her cyanotypes (item #62). It necessitated making a separately exposed print of each plate for each copy. Another disadvantage in using photographic prints for book illustration was that calotypes faded badly after a few years. Devising a means of producing permanent photographic images with the printing press was one of Talbot's chief lines of research.

In 1852 he took out his first patent for a method of printing photographic images from intaglio plates. This involved first developing an image on a light-sensitive film of bichromated gelatine spread over a steel plate. The exposed portions of the gelatine became insoluble and he washed the remainder away. He then etched the uncovered portions of the plate with a mordant to create the intaglio passages. But broad areas of etching would not hold ink when the plate was wiped, so in order to print dark intaglio tones, he had to divide them up into a cellular mosaic. However, to preserve the purity and fidelity of the photographic image, it was essential that sunlight and chemistry alone should create the printing surface; no handwork would be admissable in creating this effect. He did succeed in obtaining a grain by such devices as photographing onto the plate through a fine silk gauze. His "photo-glyphs," as he called them, were really halftone photogravures. He used the process for experimental illustrations but with unreliable results.

This diminutive pioneer example was published with a significant purpose. It shows the growth habit of the Dragon Tree of Teneriffe, believed at that time to be the oldest plant in the world. This unusual tree was illustrated first in 1810 in a lithographed drawing, and various derivative versions followed. C. P. Smyth, who knew Teneriffe well, wrote an article "On the Manner of Growth of Dracaeno Draco *in its Natural Habitat, as Illustrating Some Disputed Points in Vegetable Physiology" in the Edinburgh Botanical Society's 1860* Transactions. *He insisted that artists had grossly misrepresented its growth habit, and published this illustration to support his contention that only photography could preserve reliable records of such natural phenomena. In his view, artists tended to translate nature into the conventionalized images that European artistic training demanded, rather than to satisfy the objective needs of science.*

Smyth probably overstated his general argument, even though this particular case was proven. He also demonstrated to the Edinburgh Botanical Society that a significant new medium for botanical illustration had arrived. However, in the event, that society, and most others, continued to employ artists to prepare the great bulk of scientific botanical illustrations.

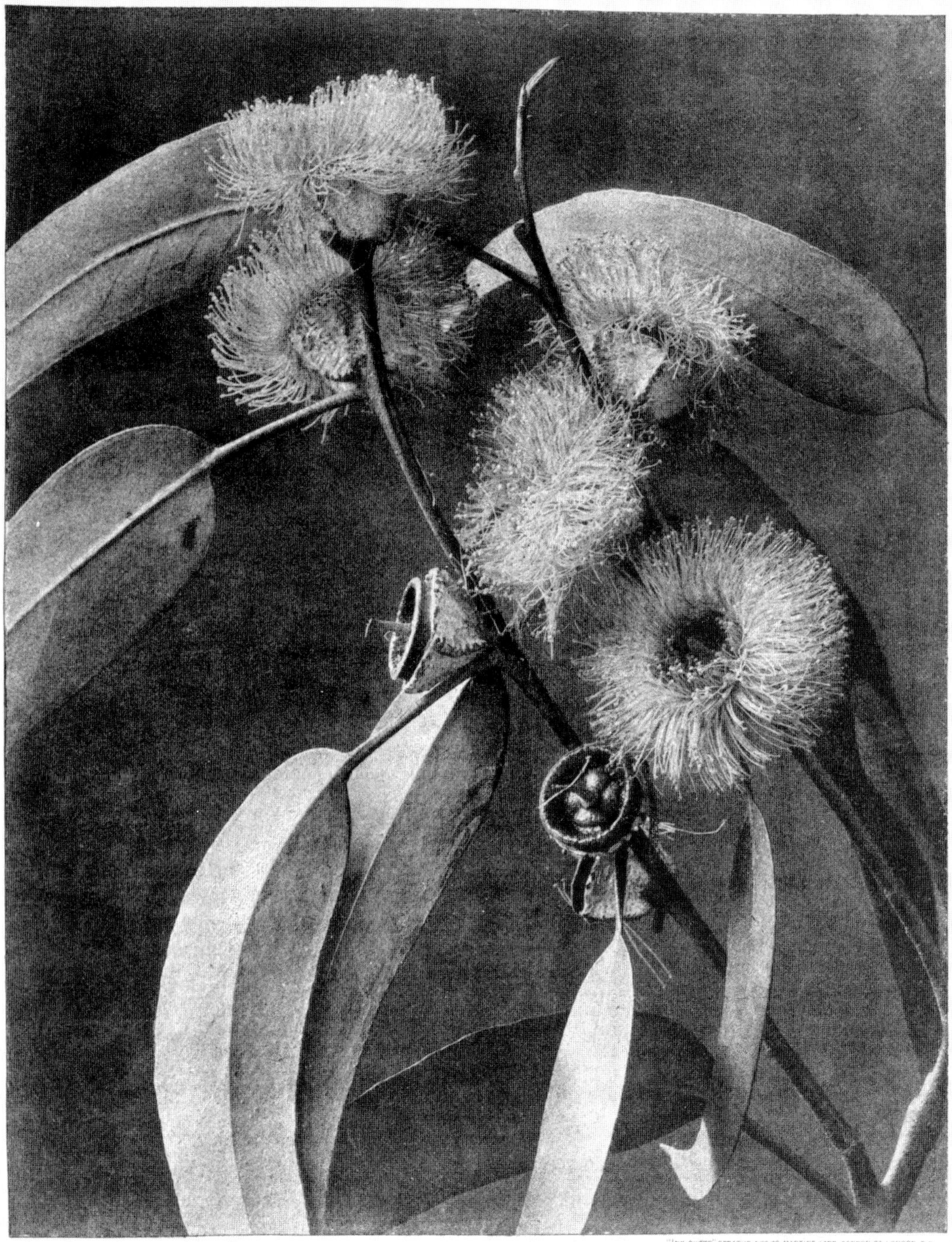

"INK-PHOTO," SPRAGUE & Co. 22, MARTINS LANE, CANNON St. LONDON, E.C.

EUCALYPTUS GLOBULUS. (BLUE GUM.)

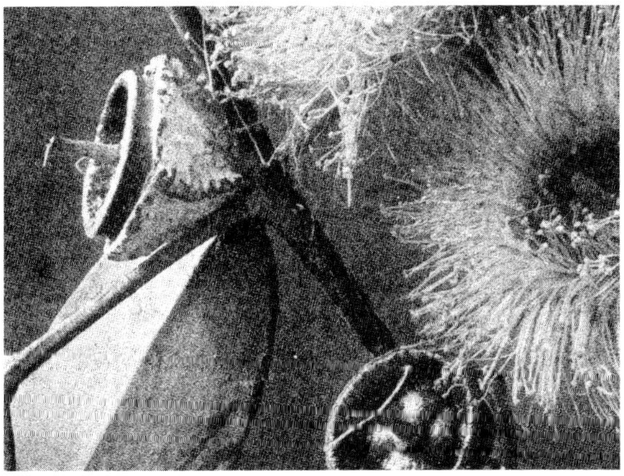

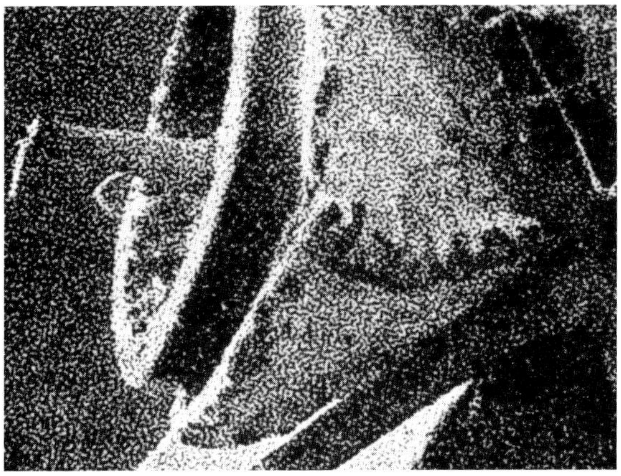

65. EUCALYPTUS GLOBULUS. (BLUE GUM). Photolithograph, collotype-grained halftone lithograph after a photograph from life ("Ink-Photo"), photographer unknown. Forming a full-page illustration printed at London in 1887 by Sprague & Co. Printed in purplish-brown ink (image area 24 x 19 cm.) on machine-made wove paper (matte finish, page size 30.2 x 22.2 cm.). In: *Gardener's Chronicle, series 3.* [Edited by M. T. Masters.] London. Volume 1, 1887. Facing page 784.

Photography gained very slow acceptance for botanical illustration. A pioneer article by Golding Bird, "On the application of heliographic or photogenic drawing to botanical purposes," published in 1839, had no obvious influence. In the 1840s, Anna Atkins prefaced her Cyanotypes of British Algae *with the explanation that "the difficulty of making accurate drawings of objects as minute as many of the Algae and Confervae, has induced me to avail myself of...cyanotype, to obtain impressions of the plants themselves." However, this apparently expressed only a personal difficulty, for other works on algae subsequently appeared with quite adequately drawn illustrations. Smyth's efforts to give photography a vital role in botanical illustration (item #64) had little effect on the preferences of botanical authors and editors in the 1860s. Botanical draughtsmen and their printmakers remained fully employed despite the existence of photography.*

Isolated books did use photography, such as S[idne]y C[ourtauld]'s Ferns of the British Isles Described and Photographed *(1877), with 20 small photographs mounted onto the plates. But successful methods of producing photographic illustrations in printer's ink came into widespread use only in the early 1880s, 40 years after photography first came to public notice.*

Among the first botanical applications of the new printing processes were some "ink photos" in the Gardener's Chronicle. *This was the trade name for a collotype-grained photolithograph. Collotype was a process for printing a tonal photographic image from a gelatin surface and worked on the same principle as lithography. A thick layer of moist photosensitive gelatine was applied to a stable substrate such as a glass plate. After exposure through a negative the gelatine was carefully heated; as it dried, its surface puckered into a reticulate pattern. The areas affected by light became less water absorbent, or else insoluble, in direct relation to exposure. Before printing, the gelatine was moistened, but the exposed image remained dry and therefore receptive to a greasy ink. The reticulate pattern of the gelatine base broke the image up into a grainy halftone. An impression of the inked image was transferred to lithographic stone for printing. Sprague & Co., the London firm that developed the process, always printed "ink photos" in brownish ink.*

The careful lighting and neutral background suggest a studio portrait, and the "ink photo" appears to give a fair reproduction of a vigorous, well defined photograph. However, the coarseness of the grain obscures details of the delicate and complex floral structures.

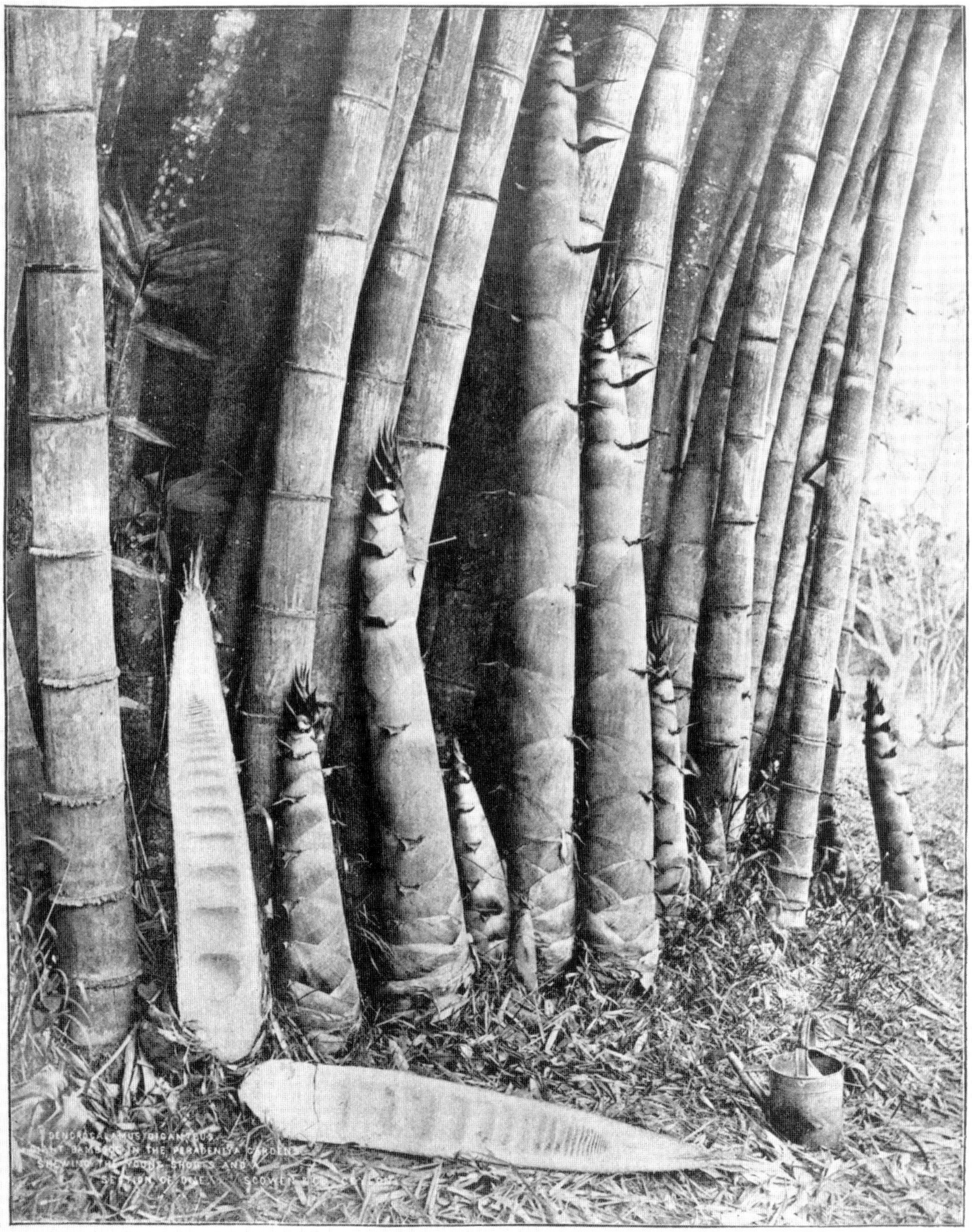

Shoots of a Gigantic Bamboo, Ceylon.

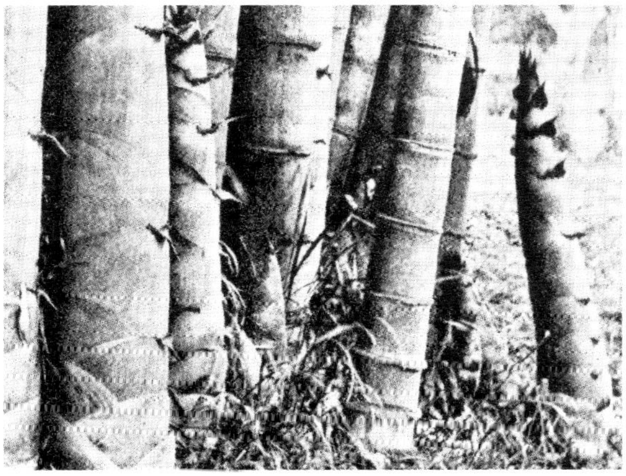

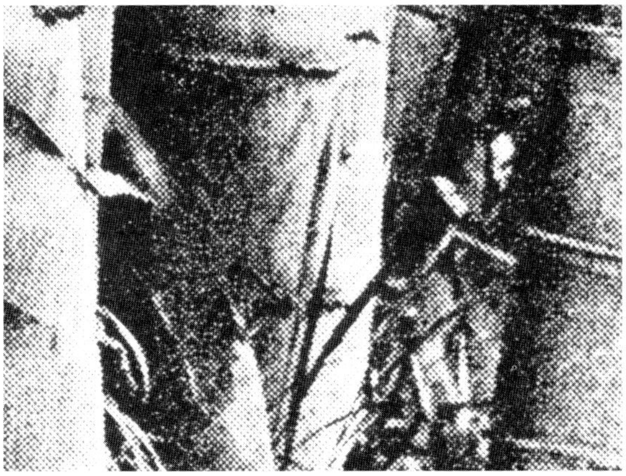

66. DENDROCALAMUS GIGANTEUS / GIANT BAMBOO IN THE PERADENIYA GARDENS / SHEWING THE YOUNG SHOOTS AND A / SECTION OF ONE. SCOWEN & CO., CEYLON. Screened halftone process engraving, etched relief block after a photograph from life by Scowen & Co., Ceylon. Forming a full-page illustration printed at London in 1892. Printed in black ink (image area 23.6 x 19 cm.) on machine-made wove paper (calendered finish, page size 29.4 x 22.5 cm.). In: *Gardener's Chronicle, series 3.* [Edited by M. T. Masters.] London. Volume 12, 1892. Facing page 308.

Smyth's insistence on photography's superiority over draughtsmanship was to acquire tacit acceptance in one respect. His criticisms were based on artists' inaccuracies in drawing the growth habit of the Dragon Tree. This was the sort of situation in which the botanical artist frequently would defer to the camera's facility in capturing a convincing image of overall plant form, habit and scale. Photography was also valuable for recording growth phases, seasonal changes, environmental associations, and so on.

In the long term the most successful means for printing photographs was the relief halftone block. The development of this process extended over several years, and by the 1880s its results reached an acceptable level of quality. A zinc block was coated with a photosensitive solution of bichromate of potassium and albumen. It was exposed to the negative through a cross-ruled screen, which broke the image up into little "cells," and the solution hardened according to the degree of exposure to light. After the unexposed gelatine had been washed away, the remaining hard image was coated with acid resist and the block was etched with a mordant. The resulting relief image was composed of dots that varied in size according to the degree of exposure.

Despite their uncompromisingly mechanical pattern, the dots were so tiny as to be scarcely visible to the naked eye, and the tones of the photograph were convincingly reproduced. Nevertheless, the depiction of very fine detail was somewhat impaired by this dot pattern. Halftone blocks were therefore suited less to precise plant portraiture than to recording broader aspects of plant form.

This plate, made with a screen of about 100 lines to the inch, shows the value of halftone photography for illustrating growth habit. At that time, good halftone blocks could only be made from "vigorous" photographs. If the photograph was too weak, a wood-engraved copy had to be substituted.

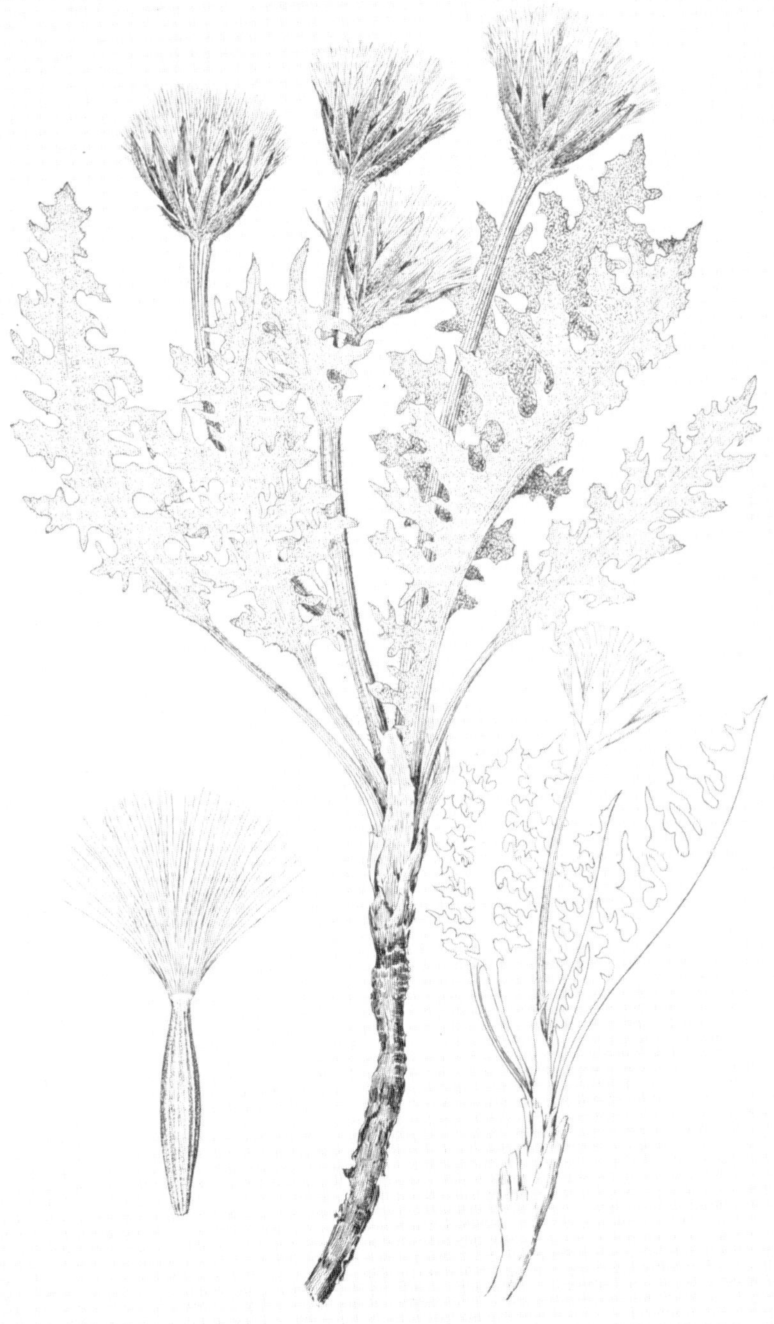

Crepis subacaulis (Kellogg.) Coville.

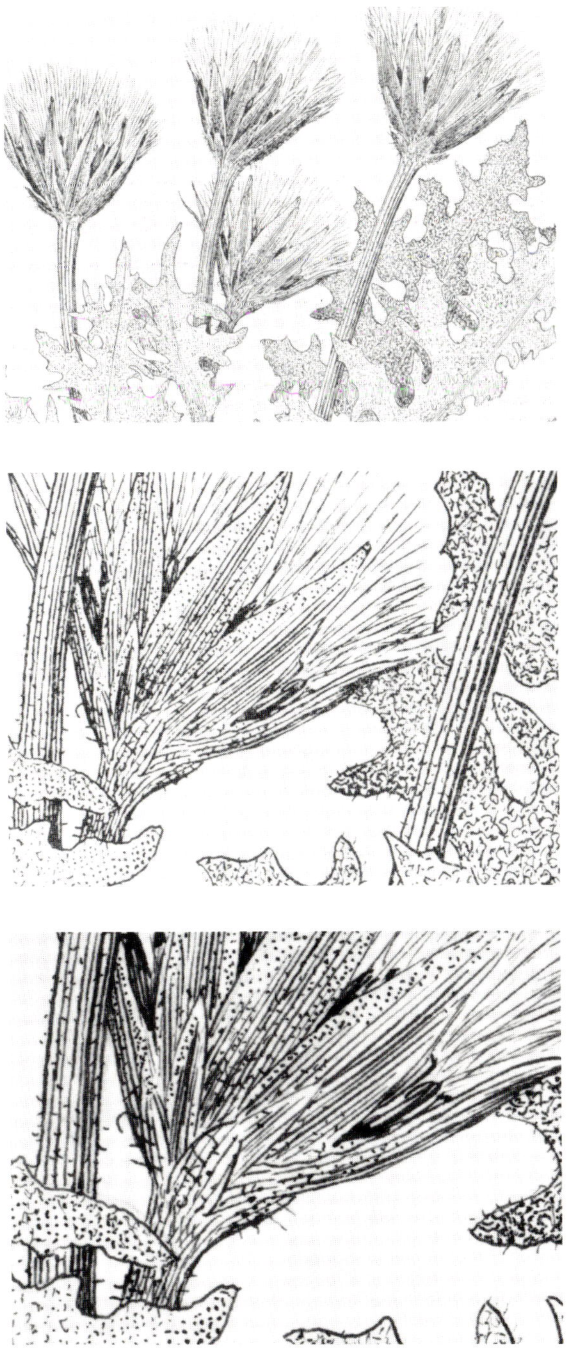

67. Crepis subacaulis (Kellogg) Coville. Zinc etching, relief "line block" after a photograph of a line drawing by an unknown American artist. Forming a full-page illustration printed at Washington in 1896 by the United States Government Printing Office. Printed in black ink (image area 18 x 11 cm.) on machine-made wove paper (calendered finish, sheet size 24.2 x 17.7 cm.). Ex: United States Department of Agriculture. *Contributions from the U. S. National Herbarium.* Washington. Volume 3, 1896. Plate 23. Shown together with the original pen-and-ink drawing (image area 25.5 x 16 cm.) done on illustration board (sheet size 35.7 x 26.5 cm.). Indefinite loan from the Smithsonian Institution, Washington.

While the printing industry was making valuable developments in the reproduction of tonal photographs, there were parallel developments in line reproduction. The etched zinc "line block" was cheap, could be printed with the letterpress, and was to have singular value in botanical illustration.

In general, botanists favored the linear style of plant portrayal and, as we have seen, a variety of successful processes had been used for printing line-drawn images. In the later 19th century, many botanical artists made their drawings in pen and ink, using the calligraphic syntax of pen to give subtle expression to their line, and employed hatched, cross-hatched or stippled shading. For generations, the reproduction of this kind of drawing had been possible only after it had been translated into a linear printmaking medium, usually by another hand. The character of the line, and thus the overall effect, was sometimes significantly altered in the process.

The new photographic methods made it possible to produce a relief block direct from the drawing with absolute fidelity to every hairline and nuance of the original. On good smooth, preferably coated, paper the "line block" could cleanly print a line as fine as 100 micrometers in width. It could not reproduce wash or pencil tones, but only work drawn in sharp, distinct lines and dots. However, this example shows that very fine stipple, or other kinds of closely laid line shading, could produce convincing tonal effects. The blockmaking process was similar in principle to that already described for halftone process engraving but, of course, it did not employ a screen.

Apart from the artist's confidence that the reader would generally see every line just as he had drawn it, there was actually a further benefit. Photography allowed the optical reduction of drawings to suit the format of the book, and the artist could turn this to his advantage. By drawing images on a much larger, more easily handled scale for reduction upon completion, he could obtain a fineness of line in the reduction that was almost beyond his power to create with the pen.

This method of illustration, with small stylistic variations, has served botanical science ever since and can be found in research literature the world over. It allows an accurate, if somewhat diagrammatic, depiction of plants that answers many of the botanist's special needs. Botanical artists were thus greatly assisted by this development in photomechanical printing. It was the trade printmaker who fell by the wayside, his craft made redundant by the camera.

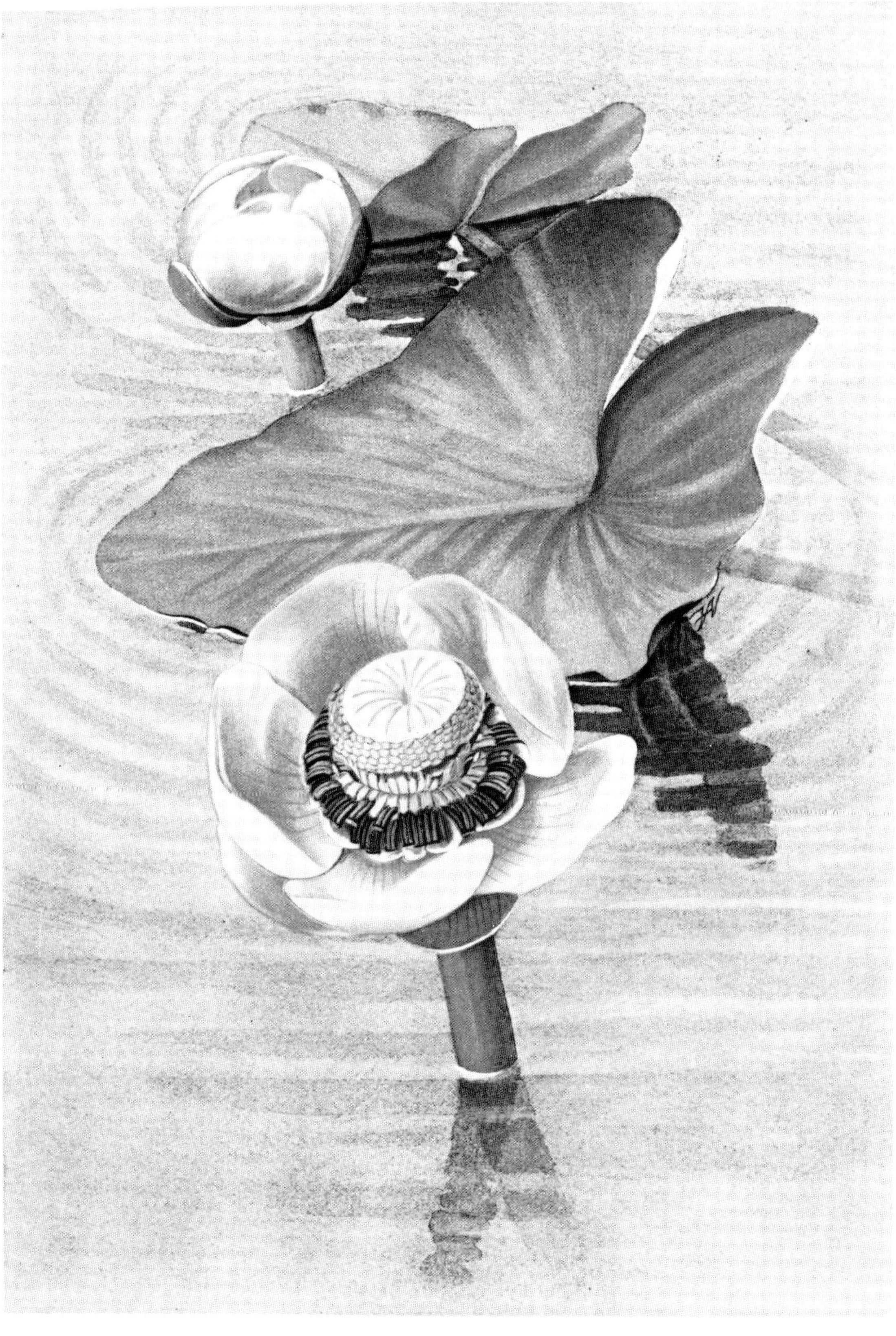

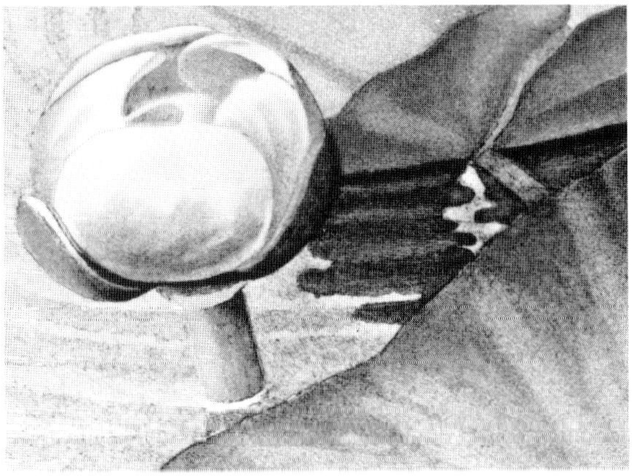

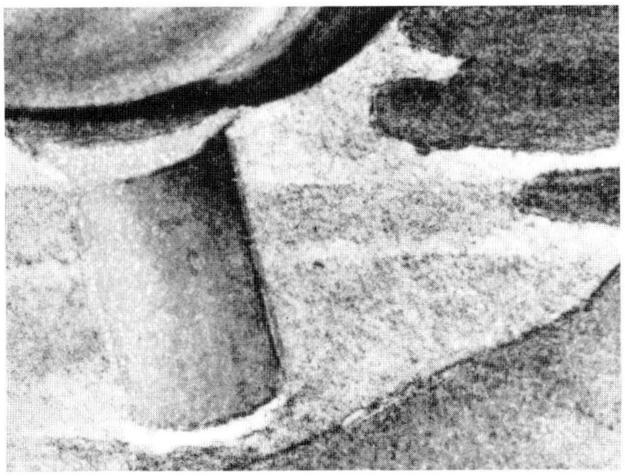

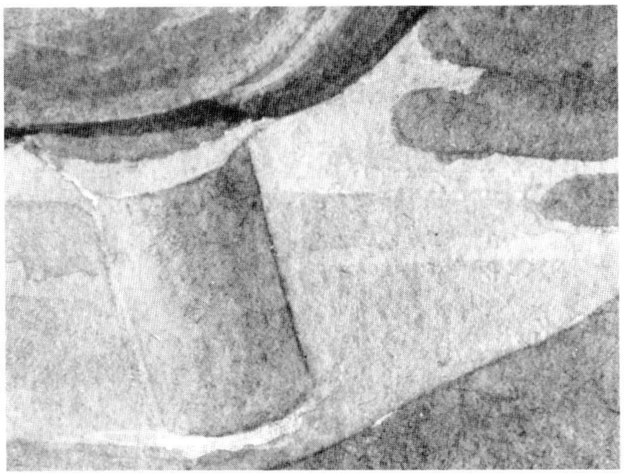

68. NYMPHAEA POLYSEPALA. Collotype, reticulate-grained photomechanical print, with six superimposed lithographic tints on color-separated stones, slightly reduced from a tinted wash drawing by Frederick Andrews Walpole (American 1861-1904). Forming a full-page illustration printed at Boston, Mass. in 1904 by the Heliotype Co. Color-printed in black, red, green, yellow, blue, grey and beige inks (image area 24.4 x 17 cm.) on machine-made wove paper (calendered finish, sheet size 30.5 x 24.2 cm.). Ex: United States National Museum. *Report...1902*. Washington, 1904. Plate 1. Shown together with Walpole's original pencil and wash drawing, tinted with opaque and transparent watercolor (image area 24.3 x 17.3 cm.), on rough-textured wove paper, mounted on board (sheet size 42 x 28.5 cm.). Indefinite loan from the Smithsonian Institution, Washington.

Superimposing a grain was an absolute necessity for photomechanical reproduction of tonal images. The cross-ruled halftone screen was a remarkably successful solution, provided one did not look too closely at the reproduction of very fine detail. The mechanical pattern of the grain was its chief shortcoming.

The collotype process, which had been applied to the "ink photo" (item #65), had none of this mechanical regularity and was susceptible of considerable refinement. By using a type of gelatine which had a greater power of absorption, and laying a thinner coat of emulsion on the substrate, it was possible to achieve a much smaller grain. The very-finest-grain collotype was virtually a continuous-tone process. In this example the grain is about three times finer than that of the "ink photo."

With its barely perceptible grain and breadth of tonal capabilities, collotype became much favored for high-quality reproductions of manuscripts, drawings and paintings. This example shows its use for reproducing a wash drawing. The extraordinary degree of fidelity to the original, shown alongside, indicates the value of this process. Unfortunately, it was more expensive to operate than the rival halftone process. Furthermore, handling the gelatine in the early stages demanded utterly vibration-free conditions and stable humidity, which lowered its practicality and thus commercial worth.

To reproduce the coloring of the original, this monochrome collotype was overprinted with six chromolithographic tints, making the admirable result something of a hybrid.

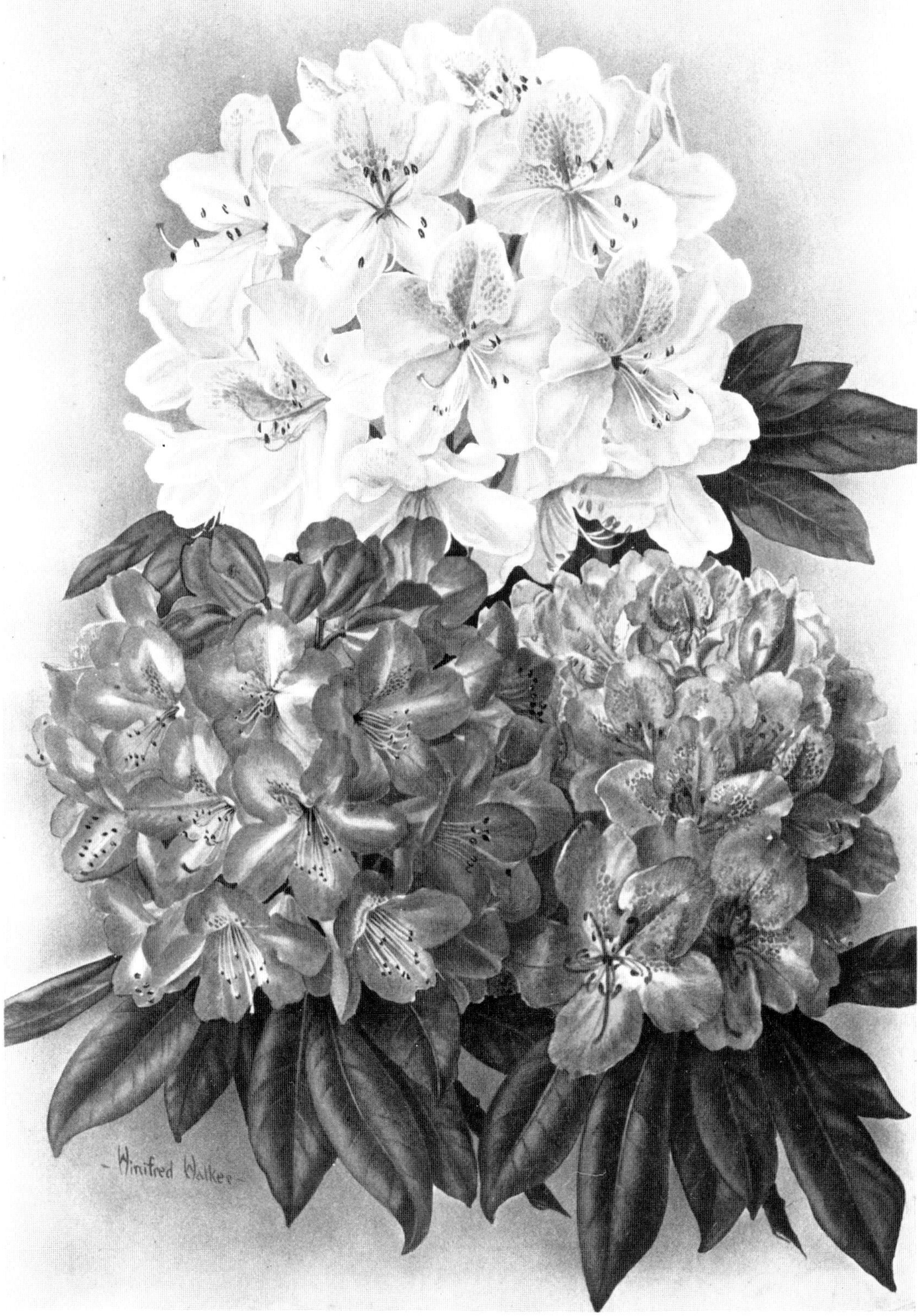
Winifred Walker

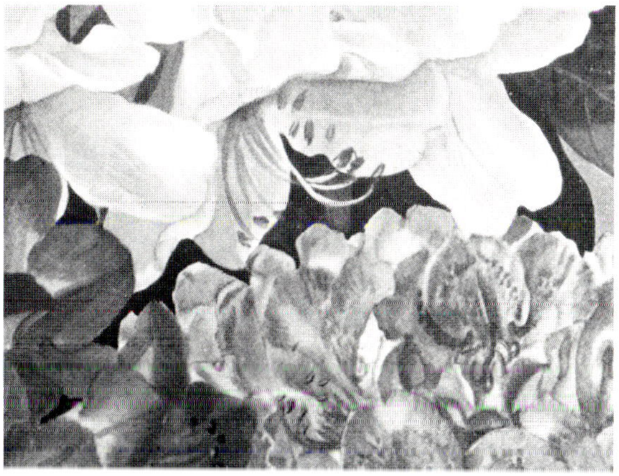

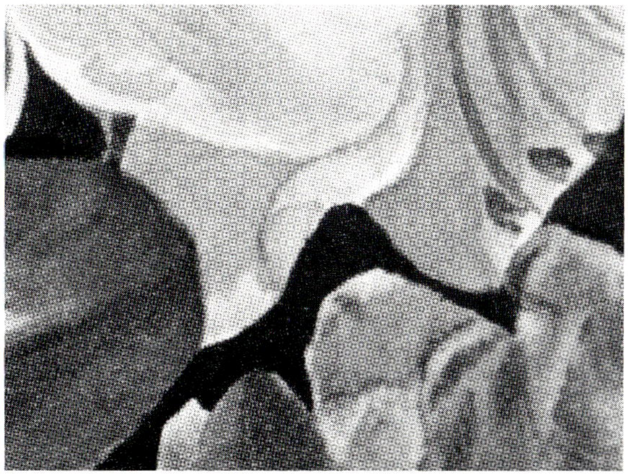

69. R. MRS. JOHN MILLAIS. / UNNAMED SEEDLING (J. Waterer). R. NELLIE MOZER. Trichromatic process engraving, relief-etched screened halftone from three color-separated transparencies, after a painting by Winifred Walker (English ca.1920). Forming a full-page illustration printed at Bushey, Herts. in 1917 by André, Sleigh & Anglo Ltd. Printed in blue, red and yellow inks (block size 29 x 20 cm.) on machine-made wove paper (calendered finish, page size 39.7 x 30.2 cm.). In: J. G. Millais. *Rhododendrons.* London, Longmans, Green & Co., 1917. Facing page 34. Nissen BBI #1369.

The impressive fidelity of the previous print was achieved by making six chromolithographic over-printings on a monochrome "key." But the color separations required much time in preparation, not to mention the skill and time needed to work them on the press. Color collotype was developed but, in the long run, it proved to be uneconomical and the results insufficiently stable for widespread commercial use. However, the monochrome capabilities of the process were unequaled, and collotype continued in use for the best facsimile work for many years.

The most successful means for photomechanical color reproduction at the turn of the century was the relief halftone. By cleverly exploiting the principle of making a whole spectrum of colors from the three primaries, it became possible to print convincing photomechanical color reproductions from only three halftone blocks. Much experimentation went into achieving this goal, and it was not until 1900 or thereabouts that trichromatic printing gained wide commercial acceptance for book illustration.

The method was fairly simple. The artwork was photographed successively through red, green and blue filters to create separate negatives for the blue, red and yellow printings. Each color block was exposed through the same ruled screen, but the screen was placed at a different angle for each separation. The aim was to orient the three meshes of dots so that when printed the dots of each grain cluster were adjacent or slightly overlapping rather than entirely coincident. The tones of each color separation were created by variations in size of the dots and the effect of the white paper showing through the interstices. With transparent, brilliantly colored inks and with dots spaced at 100 or more to the inch it was possible to deceive the naked eye into seeing a full range of colors.

Printing by this process required very precise presses to ensure that the dots lay in exactly the correct positions, but, on the credit side, only three printings were necessary. Trichromatic reproduction was widely accepted, and a number of printers added a fourth impression in black for enhanced body and richness.

This example demonstrates the capability of early pure trichromatic printing. The artist's watercolor has been successfully reproduced as a colored print entirely by photographic and mechanical means. Without the original for comparison we cannot truly judge the fidelity of this reproduction, but since it was prepared for a costly, lavishly illustrated publication we may assume that considerable care and expense were directed at obtaining good results.

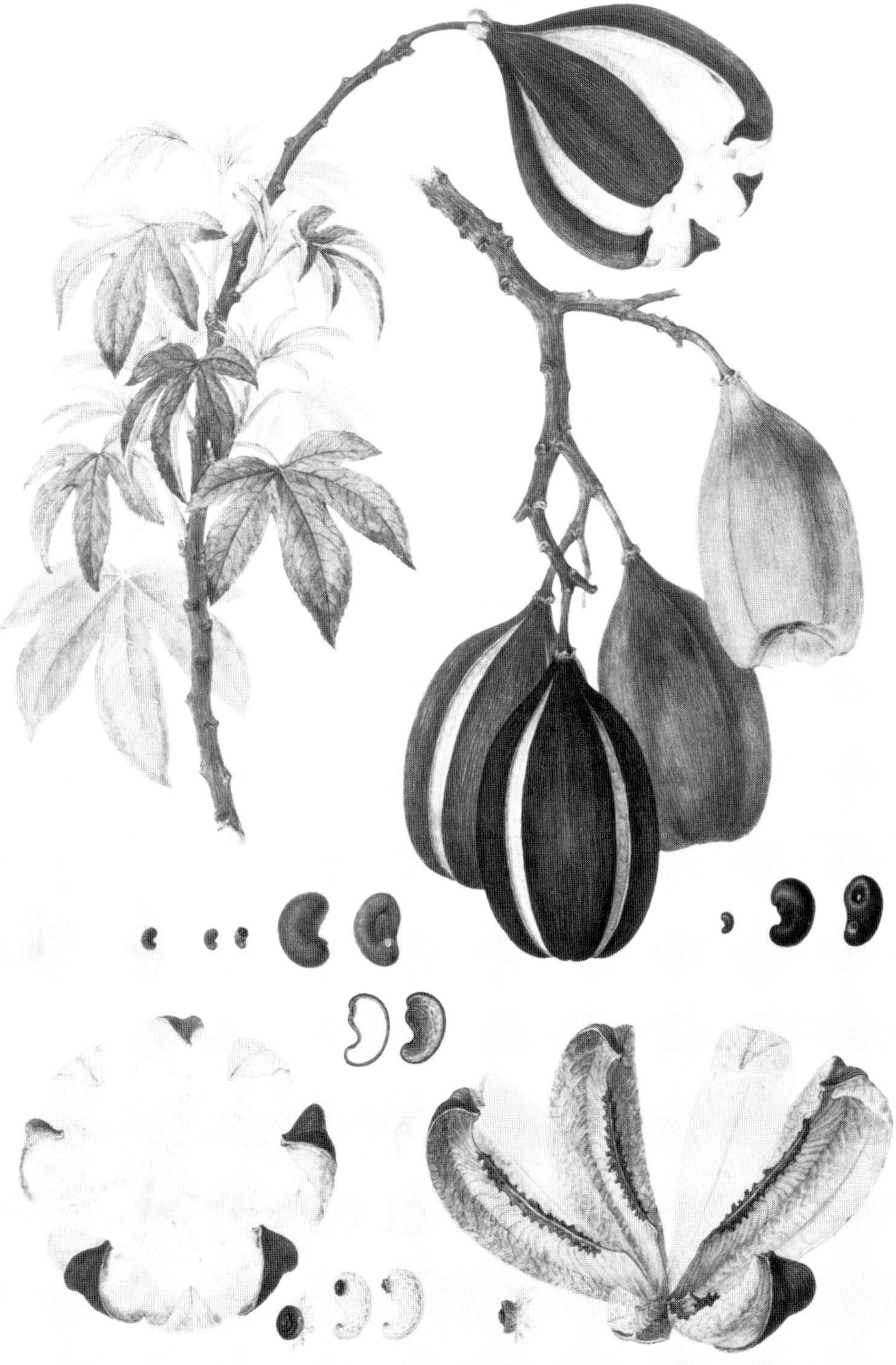

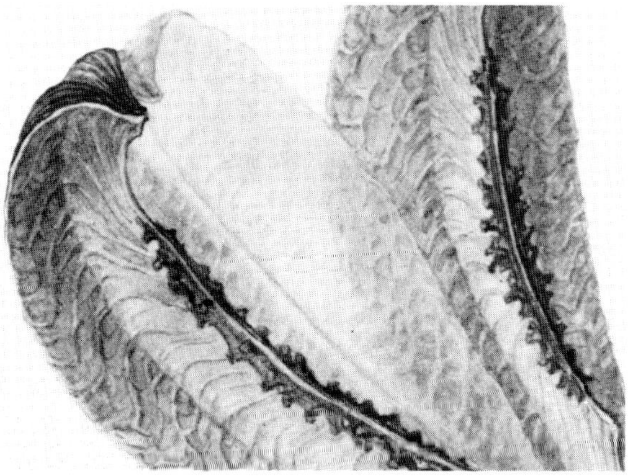

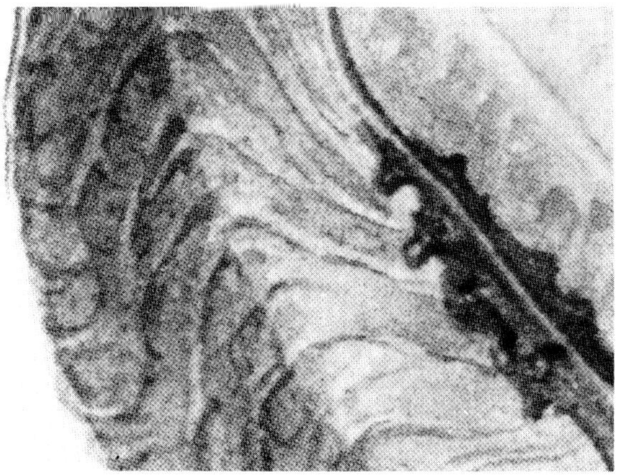

70. [Cochlospermum Gillivraei]. Photolithograph, fine-screened halftone from five color-separated transparencies, after a painting made in 1801 by Ferdinand Lukas Bauer (Austrian 1760-1826). Forming a full-page illustration printed at Bradford, Yorks. in 1976 by Lund Humphries. Color-printed by offset in black, red, yellow, blue and green inks (image area 42.5 x 30 cm.) on Inveresk Paper Co. cartridge (matte finish, sheet size 52.5 x 35.3 cm.). Proof impression prepared for: F. L. Bauer. *Australian Flower Paintings.* London, Basilisk Press, 1976. Plate 3.

The trichromatic process served well for much of the present century, and printers constantly directed their efforts at achieving closer and closer fidelity to the original color work. Screens became ever finer; inks became more vivid, reliable and amenable to printing on a wide range of papers; dot structure was modified in various ways; separation techniques became more sophisticated; and so forth. Various technical considerations led to the adoption of offset lithography in place of relief printing, and it then became possible to print on papers much more like those used by the artists themselves; coated paper gave way to matte cartridge. The principles of color printing remained the same, but when highly accurate facsimiles were demanded many of these special refinements could be employed.

In the late 18th and early 19th centuries the incomparable Bauer brothers painted hundreds of plants for their patron, Sir Joseph Banks. Both produced paintings of microscopical perfection, and it is hardly surprising that the full subtlety of their work could not be translated successfully into any print medium of the time.

Some 30 of Francis Bauer's paintings were published as Delineations of Exotick Plants...at Kew *(1796-1803), well engraved by Daniel Mackenzie and very handsomely colored. In an introductory note, W. T. Aiton, Superintendent of Kew, explained the absence of any text by saying, "...it is hoped, that every Botanist will agree, when he has examined the plates with attention, that it would have been a useless task to have compiled, and a superfluous expense to have printed, any kind of explanation concerning them; each figure is intended to answer itself every question a Botanist can wish to ask, respecting the structure of the plant it represents." Such was the level of admiration generated by Bauer's work.*

Francis relates that when Ferdinand wanted plates made for his Illustrationes Novae Hollandiae *(1813), he "could not find people capable either of engraving or coloring the plates properly, and was consequently obliged to execute every part of the work with his own hands, thus occupying far too much time." After only 15 plates had been published, Ferdinand abandoned the project.*

In 1976 it was decided to print some of these Australian paintings using the finest color printing technology of the day. The example shows the specimen plate produced for the prospectus of this costly work. Ferdinand had died 150 years earlier, but printing technology needed that much time to become capable of doing full justice to his exquisite botanical drawings.

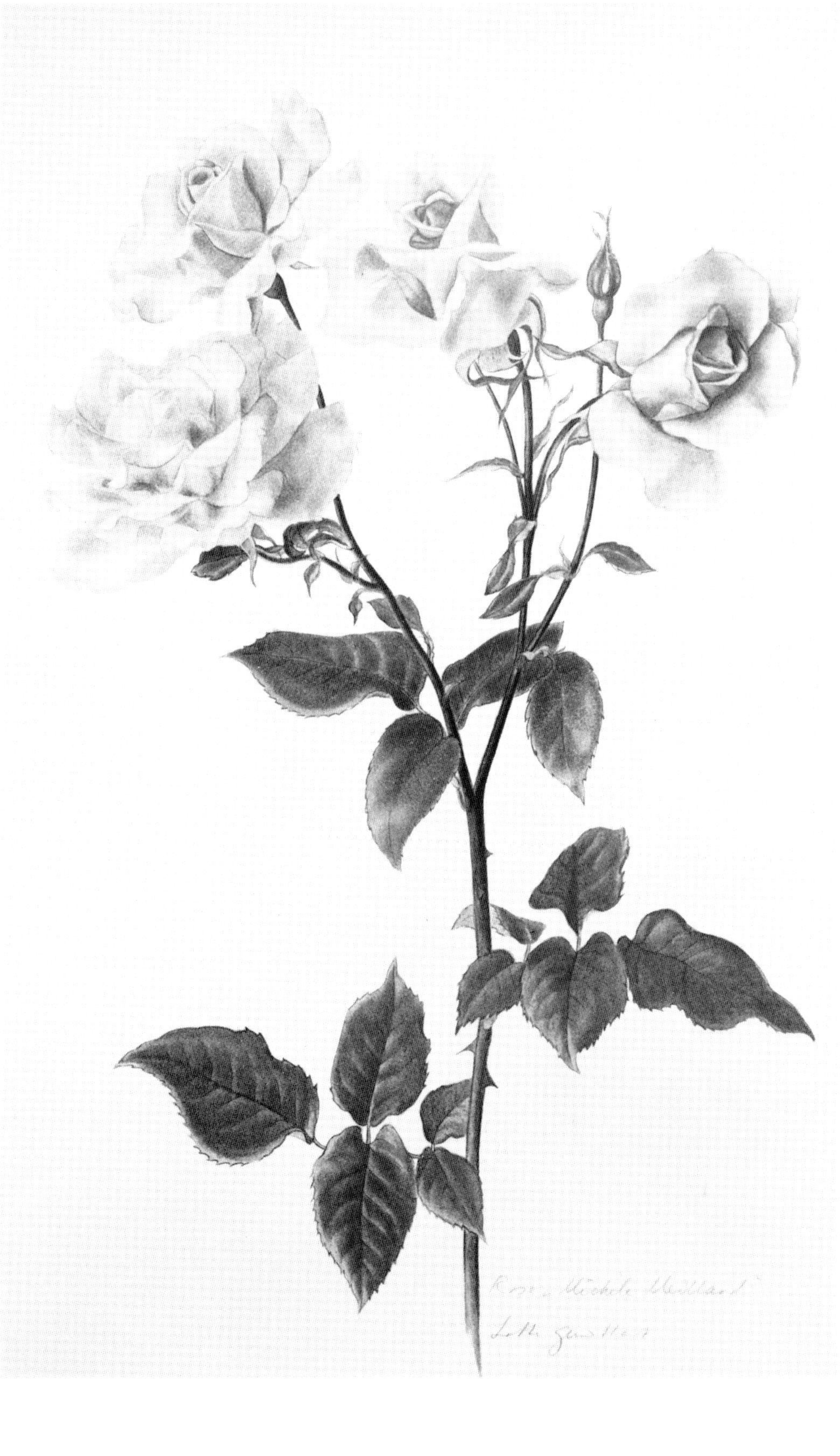

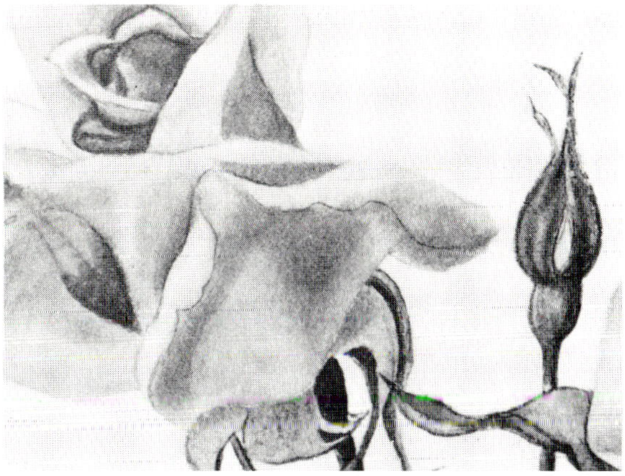

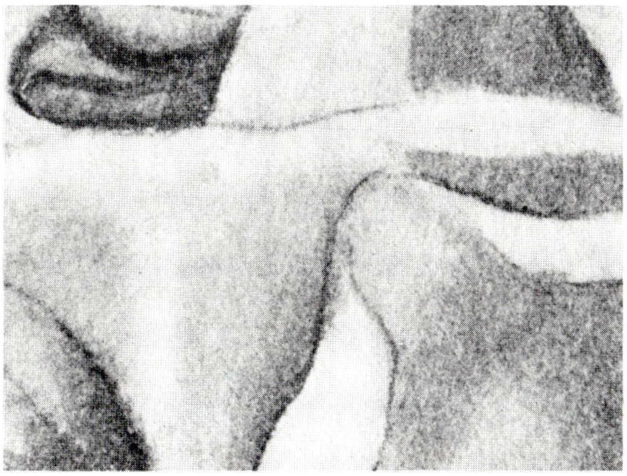

71. Rose Michèle Meilland. Offset photolithograph, collotype-grained "Granolitho" from three color-separated transparencies, after a painting by Lotte Günthart (Swiss 1914-). A separately published picture printed at Zurich in 198? by Lichtdruck/Matthieu AG. Color-printed in red, blue, yellow and black inks (image area 39 x 23.5 cm.) on machine-made paper (cream, matte finish, sheet size 41.5 x 27.8 cm.).

Two significant benefits have resulted from recent advances in the technology of color printing. First, these improvements have enabled the successful reproduction of several older, much-prized, color-plate botanical books. For example, recent years have seen good facsimile reprints of Redouté's Les Roses, *Bateman's* Orchidaceae of Mexico and Guatemala *and Lindley's* Sertum Orchidaceum. *Second, enhanced capabilities have stimulated the preparation of ambitious new large-format botanical works illustrated with reproductions of amazing fidelity. Printing technology can now reproduce almost any color artwork that an artist can create. Celia Rosser's impeccable watercolors for A. George's* The Banksias *(1982+) and Thalia Lincoln's impressively accurate crayon drawings for J. Rourke's* Mimetes *(1983) are examples of outstanding work magnificently reproduced by modern color printing.*

Although remarkably faithful facsimiles can be printed by the screened trichromatic method, there has always been some dissatisfaction with its mechanically patterned grain. Modern screens of 150 lines or more per inch produce very small dots, but the characteristic "rosette" pattern formed by the dots when they are printed at the required angles is still visible to the sharp eye.

There have recently been new efforts to eliminate the mechanically ruled screen altogether. One of these has produced a lithographic process named "Granolitho." As we saw earlier (item #68), very fine collotype grain can approximate a continuous tone. Granolitho employs a very fine collotype-grained screen for exposing each color separation. Since the grain is random, an almost unlimited number of screened colors can be overprinted without producing an irritating moiré pattern. This marriage of collotype grain with lithography harks back to the principle of the "Ink Photo" from a century earlier (item #65).

Granolitho permits finer printed definition than ruled-screen methods generally allow. On paper, the printed inks combine in much the same manner as watercolors do, making the process especially suitable for paintings done in that medium. Eight or more separations may be used, including different ones for warm and cool shades of the same color. This example shows remarkably convincing reproduction of a pure watercolor drawing, printed on a paper closely matching that used for the original.

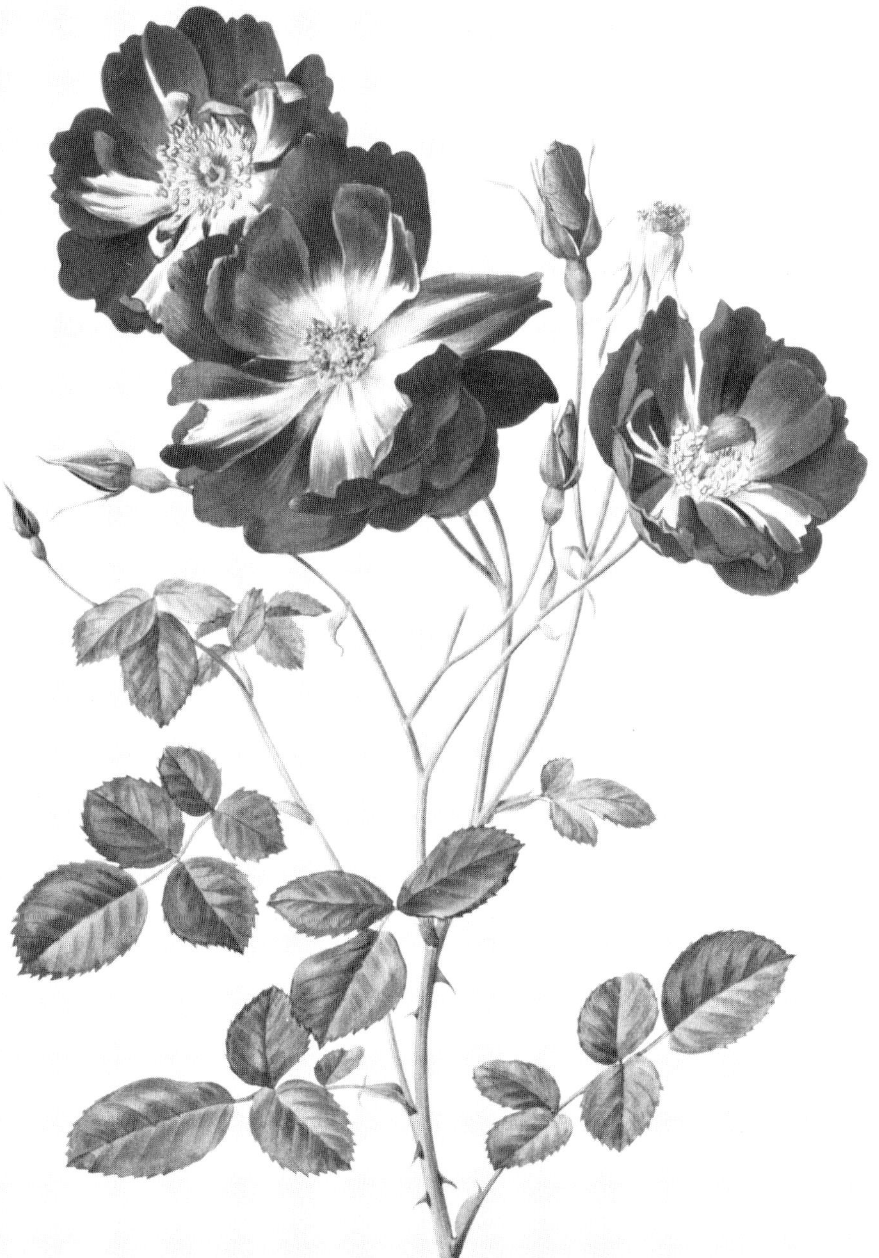

Gloire des Rosomanes

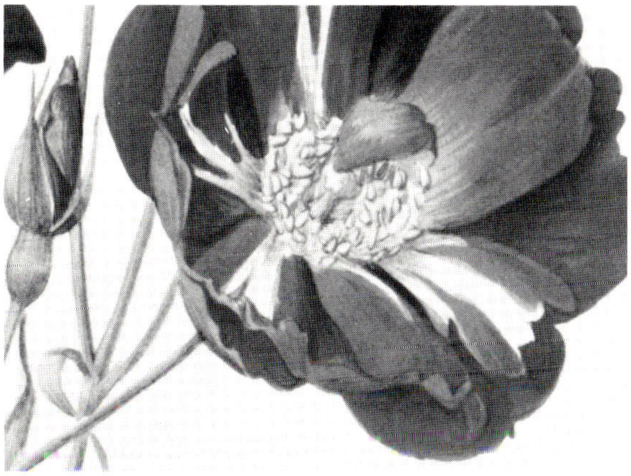

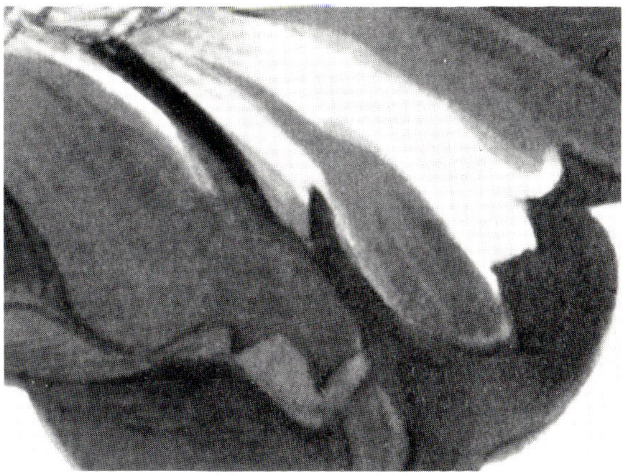

72. Gloire des Rosamanes. Photolithograph, from several screenless color-separated transparencies, after a painting by Anne Marie Trechslin (Swiss 1927-). Forming a full-page illustration printed at Dielsdorf/Zurich in 1975 by Lichtdruck SA. Color-printed in various yellow, slate blue, and red inks (image area 31 x 21.5 cm.) on machine-made wove paper (matte finish, sheet size 45.5 x 32 cm.). In: A. M. Trechslin and S. Coggiati. *Old Garden Roses.* Berne-Boll, Éditions Le Moulin SA, 1975. Plate 11.

This final example shows a brilliantly vivid watercolor reproduced by the screenless and virtually grain-free photolithographic process called "Photochrome." This technique, which dates back to the 1880s, is described by the publisher as follows: "Photochrome is a screenless flat-bed impression obtained from an aluminum plate with a special bituminous coating which is sensitive to light. The reproduction is made from grained plates and it is the graining which replaces the customary break-up of the image into minute dots of color. One of the few highly-qualified craftsmen capable of such delicate work prepares the plate for printing; the lithographer retouches by hand all the nuances of light and shade to give a precise rendering of the original. It is only by using this painstaking craftsmanship that the plates have achieved such a high degree of accord with the originals, even to the most subtle shades, so that it is virtually impossible to differentiate between them and the originals. In addition, thanks to the great number of colors used in separate printings, it has been possible to achieve the intensity of color that is vital to keep faithfully to the original."

It is indeed the intensity of the solid colors that is so striking in this print, which is absolutely free of any discernible grain. Although it is a photographically based process, Photochrome requires much skillful handwork in finishing the separations if it is to achieve its effect. But when the closest possible fidelity to an original painting is demanded, the print undoubtedly merits such costly attention to detail.

For five centuries now, the botanical artist has relied on printmakers and printers for the publication of his illustrations, with varying degrees of satisfaction. Many ingenious devices have been employed in this service and, to some extent, scientific illustration has had to accommodate their idiosyncracies. Despite admirable advances in techniques of plant photography, the artist still plays an essential rôle in botanical publication, a situation that shows no likelihood of changing. With present-day technology both the artist and the botanists he serves can publish graphic information about plants with greater accuracy and conviction than ever before. And photography does make a significant contribution too. One of the special virtues of modern printing technology is that either hand-drawn or photographic images can be mass-produced with facility and accuracy.

A SELECTION OF TYPICAL PRINTING SURFACES

Relief

A plank-grain line woodcut of [*Eruca sylvestris*] cut by Wolfgang Meyerpeck (German ca. 1560s) in 1563. Fruitwood block (22.2 x 16 cm. and ca. 2 cm. thick overall). Shown with a modern impression. Prepared for: P.A. Mattioli. *New Kreüterbuch*. Prague, G. Melantrich, 1563. Folio 195 verso. Nissen BBI #1310. For a 1565 impression see #5.

An end-grain woodblock engraving depicting a tree with a shield leaning against it engraved by Thomas Bewick (English 1753-1828) in 1782. Boxwood block (6 x 4.7 cm., and ca. 2.2 cm. thick overall). Shown with a modern impression.

A relief line photoengraving of "*Shizandra glabra*" reduced from a pen-and-ink drawing by Susan C. Smith (American 1923-) and etched by the Durham Engraving Co. in 1968. Zinc plate (18.2 x 13.7 cm.) mounted on a wood block (ca. 2.3 cm. thick overall). Shown with a modern impression. Prepared for: *Journal of the Elisha Mitchell Scientific Society* 84:354, 1968.

A relief halftone photoengraving of a photographic portrait of S. M. Tracy, made in 1903, manufacturer unknown. Copper plate (11.7 x 8.1 cm.) mounted on a wood block (ca. 2.3 cm. thick overall). Shown with a modern impression. Prepared for: *Journal of Mycology* 9: facing p. 81, 1903.

Intaglio

An intaglio line- and stipple-engraved plate of "*Geometra Piniaria*" engraved and etched by John Curtis (English 1791-1862) in April 1812. Copperplate (20.4 x 14.3 cm., and ca. 1.5 mm. thick). Shown with a modern impression. Lent by Francis M. Hueber, Chevy Chase, Maryland.

An intaglio aquatint-etched plate entitled "Old House near Boulogne," aquatinted by James David Smillie (American 1833-1909) in the early 19th century. Copperplate (17.8 x 12.9 cm., and ca. 1.5 mm. thick). Lent by the Division of Graphic Arts, National Museum of American History, Smithsonian Institution.

Planographic

Planographic drawings of "Pl. 20. ACER RUBRUM. Red Maple." and "Pl. 25. ACER SPICATUM. Mountain Maple.," drawn on stone by Joseph Prestele (German/American 1796-1867) after drawings by Isaac Sprague (American 1811-1895) ca. 1850. Lithographic limestone (36 x 51.5 cm., and ca. 5.3 cm. thick). Shown with a contemporary impression of "Pl. 25." (The reverse of the stone also has two drawings: "Pl. 39. CERCIS CANADENSIS. Red Bud." and "Pl. 46. PRUNUS AMERICANA. Wild Plum.") Prepared for: A. Gray. *Plates Prepared Between the Years 1849 and 1850, to Accompany a Report on the Forest Trees of North America*. Washington, Smithsonian Institution, 1891. Nissen BBI #751. Lent by the Division of Graphic Arts, National Museum of American History, Smithsonian Institution.

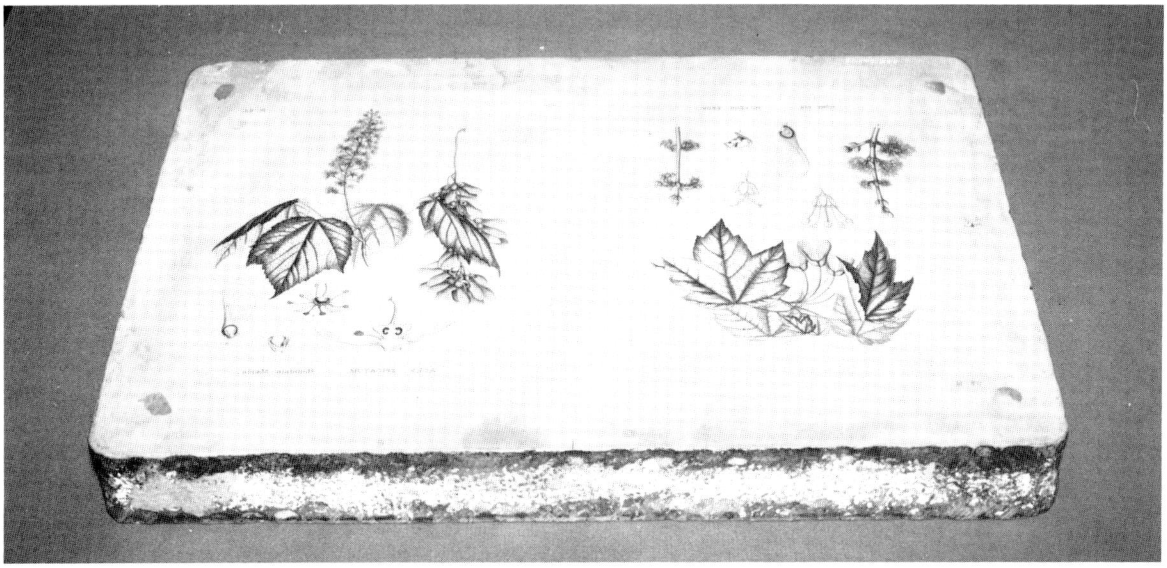

Sources of Information

Arber, A. 1938. *Herbals: Their Origin and Evolution. A Chapter in the History of Botany 1470-1670.* Second edition, rewritten and enlarged. Cambridge, Cambridge University Press. xxiv + 326 pp. Reprinted 1963.

Bayard, J. and E. D'Oench. 1976. *Darkness into Light: The Early Mezzotint.* [New Haven], Yale University Art Gallery. 30 pp.

Blunt, W. 1955. *The Art of Botanical Illustration.* Third edition. London, Collins. xxxii + 304 pp.

_____. 1963. The Illustrators of "Sertum Anglicum." In: L'Heritier de Brutelle, C.-L. *Sertum Anglicum 1788. Facsimile with Critical Studies and a Translation,* pp. lxi-1xvi. Pittsburgh, Hunt Botanical Library.

_____ and S. Raphael. [1979.] *The Illustrated Herbal.* London, Frances Lincoln. 191 pp.

Bridson, G. D. R. 1976. The Treatment of Plates in Bibliographical Description. *Journal of the Society for the Bibliography of Natural History* 7: 469-488.

Brunner, F. 1964. *A Handbook of Graphic Reproduction Processes.* Second edition. Teufen, A. Niggle. 379 pp.

Burch, R. M. 1910. *Colour Printing and Colour Printers.* Second edition. London, Sir Isaac Pitman and Sons. xviii + 281 pp.

Cave, R. and G. Wakeman. 1967. *Typographia Naturalis.* Wymondham, Brewhouse Press. 36 pp.

Church, A. H. 1919. Brunfels and Fuchs. *Journal of Botany 57: 233-244.*

Curtis, W. H. 1941. *William Curtis 1746-1799.* Winchester, Warren and Son. xvii + 142 pp.

Dunthorne, G. 1938. *Flower and Fruit Prints of the 18th and Early 19th Centuries. Their History, Makers and Uses, with a Catalogue Raisonné of the Works in Which They are Found.* Washington, The Author. xiv + 275 pp. illus. Reprinted London, Holland Press, 1970.

Getscher, R. H. 1980. *A Garden of Prints: An Exhibition of 18th and 19th Century Botanical Illustrations. Fine Arts Gallery, Grasselli Library, John Carroll University, September 26-October 17, 1980.* Cleveland. 102 pp.

Harris, E. M. 1968-1970. Experimental Graphic Processes in England, 1800-1859. [Parts] 1 (-V). *Journal of the Printing Historical Society* 4: 33-86; 5: 41-80; 6: 53-89.

Henrey, B. 1975. *British Botanical and Horticultural Literature Before 1800.* London, Oxford University Press. Three volumes.

Hullmandel, C. 1824. *The Art of Drawing on Stone, Giving a Full Explanation of the Various Styles of the Different Methods to be Employed to Ensure Success, and of the Modes of Correcting, as well as the Several Causes of Failure.* London, C. Hullmandel and R. Ackermann. vii + 19 pp., 19 pls.

Hunnisett, B. 1980. *Steel-Engraved Book Illustration in England.* London, Scolar Press. xvi + 263 pp.

Hunt Botanical Library. 1958-1961. *Catalogue of Botanical Books in the Collection of Rachel McMasters Miller Hunt.* [Compiled by] J. Quinby and A. Stevenson. Pittsburgh. Two volumes in three.

Kainen, J. 1959. Why Bewick Succeeded: A Note in the History of Wood Engraving. *Bulletin of the United States National Museum* 218: 186-201.

Knight, D. M. 1977. A Note on Sumptuous Natural Histories. *Annals of Science* 34: 311-314.

Muir, P. H. 1971. *Victorian Illustrated Books.* London, B. T. Batsford. xv + 287 pp.

Müller, H. E. 1985. Granolitho—A New Printing Technique. In: Günthart, L. *Linger Golden Light,* pp. 223-225. Pittsburgh, Hunt Institute for Botanical Documentation.

Nissen, C. 1966. *Die botanische Buchillustration: Ihre Geschichte und Bibliographie.* Zweiter Auflage. Durchgesehener und verbesserter Abdruck der zweibändigen Erstauflage ergänzt durch ein Supplement. Stuttgart, A. Hiersemann. x + 264 + [iv] + 316 + 97 pp.

Salaman, M. C. 1906. *The Old Engravers of England.* London, Cassell and Co. viii + 224 pp.

Savage, S. 1923. The "Hortus Floridus" of Crispijn van de Passe the Younger. *Transactions of the Bibliographical Society, series 2, 4: 181-206.*

_____. 1923. Early Botanic Painters. *Gardeners' Chronicle, series 3,* 73: 92-93, 148-149, 200-201, 336-337.

Schaaf, L. 1981. Piazzi Smyth at Teneriffe: Part 2, Photography and the Disciples of Constable and Harding. *History of Photography* 5: 27-50.

_____. 1985. *Sun gardens: Victorian Photographs by Anna Atkins.* New York, Aperture. 104 pp.

Schnauss, J. [1889.] *Collotype and Photo-Lithography.* London, Iliffe and Son. 170 pp.

Stafleu, F. A. 1963. Redouté—Peintre de Fleurs. In: *A Catalogue of Redoutéana Exhibited at the Hunt Botanical Library 21 April to 1 August 1963,* pp. 1-32. Pittsburgh, Hunt Botanical Library.

_____ and R. S. Cowan. 1976 + . *Taxonomic Literature: A Selective Guide to Botanical Publications and Collections with Dates, Commentaries and Types.* Second edition. Utrecht, Bohn, Scheltema and Holkema. Volume 1 + . (In progress.)

Twyman, M. 1970. *Lithography 1800-1850.* London, Oxford University Press. xxi + 302 pp.

Wakeman, G. 1973. *Victorian Book Illustration: The Technical Revolution.* Newton Abbot, David and Charles. 182 pp.

Widder, F. 1960-1963. *Erzherzog Johann und seine Icones Plantarum.* Munich, Verein zur Schutze der Alpenpflanzen und -Tiere. 32 pp. ("Sounderdruck aus dem *Jubiläumsjahrbuch 1900-1960*, 25. Band, S. 104-117 (1960) und aus dem *Jahrbuch*, 28. Band, S. 16-33 (1963) des Vereins zum Schulze der Alpenpflanzen und -Tiere e. V.")

Wilkinson, W. T. 1887. *Photo-Engraving and Photo-Lithography in Line and in Half-Tone, also Collotype and Heliotype.* Second edition. London, England Bros. vii + 197 pp.

Index

References are by catalogue entry number.

Set in Bookman phototype by Mangis & Associates

Printed offset on Cameo Dull 80# text and cover by Hoechstetter Printing Company, Inc.

Halftones and color separations by Hoechstetter Printing Company, Inc.

Reproduction photography by Frank A. Reynolds

Designed by Rob Roy Kelly with the assistance of Richard Weber, Frank Longo and Philip Sones

Graphics coordination by The Waterkotte Co.